DATE DUE

MR 29 '99			
MR 14 00			

DEMCO 38-296

ABOUT THE ADOPT-A-STREAM FOUNDATION

The Adopt-A-Stream Foundation (AASF) was established in 1985 to promote environmental education and stream restoration. The long term goal of the AASF is to see *all* streams benefiting from good stewardship by the people that live, do business, and recreate in the watersheds surrounding those streams.

This goal will not be achieved until everyone has a good understanding of the functions and values of streams. To address this situation, the AASF has a small professional staff to conduct "Adopt-A-Stream," "Streamkeeper Field Training," and "Watershed Education" workshops for teachers and community group leaders. Each participant in these events is encouraged to share their new knowledge with fellow watershed residents, and to organize watershed-wide stream protection and enhancement projects.

The AASF Board of Directors also recognizes a growing need to develop environmental educational materials that encourage all of the "nonparticipants" to become part of the solution to the problems that our streams are facing today. In addition to the training workshops mentioned above, the AASF has produced *Adopting A Stream: A Northwest Handbook* and *Adopting A Wetland: A Northwest Guide*. The *Streamkeeper's Field Guide* and *The Streamkeeper* video are the newest educational tools that are designed to stimulate and guide you and your friends and neighbors into action.

THE
Streamkeeper's
Field Guide

■

WATERSHED INVENTORY AND STREAM MONITORING METHODS

■

BY TOM MURDOCH AND MARTHA CHEO

with KATE O'LAUGHLIN

ILLUSTRATED BY THOMAS WHITTEMORE

WITH GARY LARSON, DAVE HORSEY, STEVE GREENBERG,

CHRIS BRITT, BRIAN BASSET, KEN ALEXANDER, TOM TOLES

A PRODUCT OF THE ADOPT-A-STREAM FOUNDATION

LIBRARY OF CONGRESS CATALOGING-IN-PUBLICATION DATA

Copyright 1991, 1996 by the Adopt-A-Stream Foundation. All rights reserved. Except for pages

this book may be reproduced in any form, except for brief
sion from the publisher.

on how streams and their surrounding watersheds function,
nventory and stream monitoring for volunteers, tips on
t Streamkeepers putting watershed inventory and stream
the protection and restoration of our nation's streams.

ISBN: 0-9652109-0-1

Printed in the United States of America

Principal Authors:
Tom Murdoch, Martha Cheo

Contributing Writer:
Kate O'Laughlin

Graphic Design, Typesetting, Editing:
Sidnee Wheelwright, T.R. Morris, Tim Northern

Published by:
The Adopt-A-Stream Foundation
600-128th Street SE
Everett, WA 98208
(206) 316-8592

Printed with soy-based inks

ABOUT THIS PUBLICATION

This book is really the fifth edition. It is based on the 110-page publication produced by the Adopt-A-Stream Foundation in 1991 with grants from the Washington State Department of Ecology, the US Environmental Protection Agency, Snohomish County (WA) Planning Department, and the Bullitt Foundation. The Adopt-A-Stream Foundation is solely responsible for the production of this edition, which has grown considerably in breadth as a result of extensive field testing and contains many "thought provoking" illustrations.

Table of Contents

Foreword

WHAT IS A STREAMKEEPER?

The term evolved from the English "River-keepers" in the 1700's. River Keepers protected the habitat of salmon and trout streams. They "preserved an unimpeded flow of water, prevented poaching, encouraged the breeding of fly (food sources for salmon and trout), and especially, to see the weeds along the banks in which the fly naturally breed were cared for properly." *River-keeper*, J.W. Hill, 1934.

Modern Streamkeepers have wider responsibilities. They examine closely the physical, chemical, and biological nature of their streams. They test the water quality, survey fish and wildlife habitat, and keep abreast of land uses affecting entire watersheds. Streamkeepers know that the health of their streams is dependent on what we humans do to the landscape that surrounds their streams.

Streamkeepers also know that the real resource managers are you, your friends and neighbors, and those people you elect to government offices at the local level. Collectively you make decisions every day that affect our environment, and subsequently our water resources. While there are lots of scientists and biologists in numerous agencies concerned about the environment, they often are only in the position to give technical advice.

By becoming a Streamkeeper and monitoring the health of your stream, you become an important player in your stream's future condition. You can become an extra set of eyes and ears that government agencies need to ensure that your stream iS looked after on a regular basis. Your inventory and monitoring data can serve as the basis for protecting your stream from potenially harmful land use decisions, or restoring your stream if it is already degraded.

The Adopt-A-Stream Foundation was established to promote stream stewardship and environmental education. The Foundation's long term goal is to have all of us "adopt" streams....in other words, to become Streamkeepers. There are many Streamkeepers that are helping us reach this goal. Some of their stories are highlighted in Chapter 10, "Streamkeeper Tales."

ANYONE CAN BE A STREAMKEEPER

The real key to becoming a successful Streamkeeper is to get others in your watershed to become Streamkeepers as well. You may have taken a quick glimpse below of the typical parts and tools of a Streamkeeper and found a few items that you don't have or are not working quite as well as they used to. Don't despair, for you can find those missing parts or tools by seeking out friends and neighbors, schools, community groups, and local government representatives to join in your Streamkeeping effort.

A person in a wheelchair may not be able to walk the stream and collect field data, but she or he can record the information, keep the records, do research, and work a telephone. A composite Streamkeeper team of senior citizens can do it all. A first grader can take the temperature of a stream as well as a Ph.D . . . the point being that you do not have to be a "gortex clad yuppie" with a college degree to be a Streamkeeper.

Let's face it: If we are going to take care of our streams successfully, it is up to all of us humans, no matter what our age, background, or physical capabilities.

STREAMKEEPER PARTS

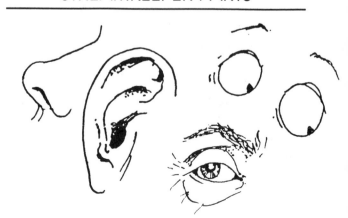

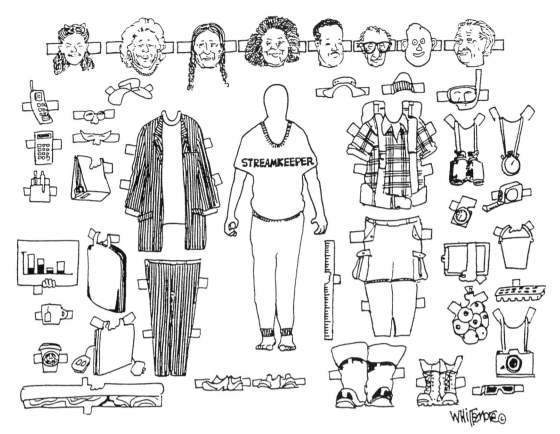

WHY BECOME A STREAMKEEPER?

Streams and rivers are invaluable. They provide vital habitat and a source of fresh water for fish and wildlife. Streams also provide humans with water for drinking, irrigation, industrial supplies, power production, transportation, flood control, fishing, boating, swimming and aesthetic enjoyment.

Serving so many human uses has led to the modification of streams and their surrounding watersheds. Most of our streams and rivers have been dammed to create reservoirs and provide power. We have channelized, dredged, and levied our streams in often ill-fated attempts to control flooding. Removing streamside vegetation and constructing impervious surfaces, like parking lots and streets, has increased runoff, slowed ground water recharge, and added a variety of pollutants from automobiles and the by-products of our daily lives.

Collectively, we humans are the primary cause of degradation to stream systems. But we can also create the solutions! We can be instru-mental in protecting and restoring fish populations, wildlife habitat, and water quality. Streamkeepers can also help ensure that future generations can enjoy the beauty and recreational opportunities our watersheds have to offer. We hope that this book will help you feel empowered to take action and become a Streamkeeper.

How To Use This Book

For those millions of you who are intrigued by the wonders of streams, and want to find out how they really function, *The Streamkeeper's Field Guide* is for you. This book will help you gain a thorough understanding of the relationship between natural factors, land use impacts, and your stream's health. It provides methods for obtaining a holistic picture of your stream's watershed, as well as directions on how to collect detailed information about your stream. *The Streamkeeper's Field Guide* also illustrates how you can put the information you collect to use, ensuring that your stream's future is bright.

The Streamkeeper's Field Guide starts you out in Chapters 1 and 2 with the big picture - understanding what's going on in your stream's entire watershed, from headwaters to mouth. It then leads you to looking at things on smaller and smaller scales. After an introduction to stream monitoring in Chapter 3, Chapter 4 shows you how to get a sense of the condition of your stream by looking at one or a few representative measured sections of your stream, called **stream reaches**.

Once you have mapped and surveyed your selected stream reach(es), you have many options for getting to know your stream in more detail. Chapter 5 covers methods to measure physical aspects of your stream, such as cross section shape, stream bottom substrate, and flow. Chapter 6 introduces benthic macroinvertebrates (spineless critters that live on the stream bottom) and how to collect and use them as indicators of stream health. Chapter 7 covers the more traditional water quality tests involving the chemical character of the stream water, such as pH and dissolved oxygen.

The types of watershed inventory and stream monitoring techniques presented in *The Streamkeeper's Field Guide* are fairly simple, inexpensive, and can be accomplished with easily obtainable or homemade equipment. Although they may be less "high-tech" or quantitative than methods used by agency scientists, they can still provide you with valuable information. Streamkeepers can play an important role in screening a watershed for potential problems that agency staff can explore in more detail if necessary.

THE FAR SIDE   By GARY LARSON

But it's not always necessary to bring in the agency scientists for verification. Streamkeepers can and have produced data that is at the same technical level as that of government agencies. Chapter 8 offers guidelines to help you ensure that your observations and conclusions are viewed as credible, even with simple, homemade methods. As long as you follow uniform procedures to assure and maintain the quality of your information, your efforts can produce valuable, usable data. By following the guidelines offered in Chapter 8, you may find that agency representatives will be calling *you* for assistance and counsel!

In addition to making sure your data is credible, you will also need to know how to present your information effectively if you want it to be considered in land use decisions.

You may need to share the information you gather with your neighbors, other watershed residents, or with local government elected officials. Chapter 9 provides basic tips on presenting the information you collect so you'll be able to carry the day in that all important public hearing on a land use that may determine your stream's future health.

And finally, Chapter 10 will give you insight into putting all of your information to use. Tips provided range from how to organize a stream clean up, to how to influence land use decision makers. This chapter illustrates that Streamkeepers have the power to take action to protect or restore streams, and encourage others to make environmentally sound decisions.

Interested? Then jump in and get your feet wet!

Acknowledgments

Between October 1991 and November of 1995, the Adopt-A-Stream Foundation (AASF) went on an odyssey around the Pacific Northwest. Our purpose was to train volunteers how to adopt streams.

During that time we journeyed to watersheds of all types. We investigated pristine mountain streams, and streams channelized through feed lots. We saw streams teeming with trout and salmon, and we saw areas totally devoid of fish and wildlife. Some of the watersheds observed flowed through timber harvest areas, agriculture fields, and shopping malls.

During those trips we had the good fortune to meet and teach more than 4500 dedicated volunteers how to go about "adopting" or becoming stewards of their streams and surrounding watersheds.

We also researched environmental education programs across 39 US states and British Columbia, evaluating different stream monitoring and watershed inventory procedures. That research, and participation in the development of the Environmental Protection Agency's *Streamwalk Program*, led the AASF to develop the *Streamkeeper's Field Guide*. We wanted to encourage the public to look beyond the stream corridor, to that large area around the stream -- the watershed.

The *Streamkeeper's Field Guide* has been field tested by 555 teachers, community group leaders and dedicated volunteers who participated in a series of intensive "Streamkeeper Field Training" workshops conducted by the AASF. We are happy to report that many of these Streamkeepers are now training students and members of their communities to become Streamkeepers. These Streamkeepers are causing a "ripple effect" in their watersheds.

As you can see, *The Streamkeeper's Field Guide* is the product of a team effort. The principal authors of the book are Tom Murdoch, Martha Cheo, and Kate O'Laughlin. Kate is a former

AASF Ecologist who currently heads an ambient surface water monitoring program for King County, Washington. Martha, a past AASF Ecologist, is now an instructor at the Institute for Ecosystem Studies in New York. Tom is the AASF founder and director. Sidnee Wheelwright and T.R. Morris guided the book's final layout. The book could not have been produced, however, without the help provided by present and past AASF staff members Jo Goeldner, Kathy Koss, David Whiting and Tim Northern who, in addition to editorial assistance, contributed field experience and ideas that give this book some of its unique character.

The AASF is particularly grateful to the teachers and volunteers who served on our Citizen's Advisory Committee to ensure that the book was as user friendly as possible.

We are also grateful to the team of technical advisors from resource management agencies who ensured that the book is technically accurate.

Eleanor Ely, Editor of the *Volunteer Monitor*, contributed a helpful article for the section on credible data.

Seattle cartoonist Thomas Whittemore put together a great supporting cast of artists and editorial cartoonists who make this book come alive. Thomas, along with syndicated editorial cartoonists Ken Alexander and Brian Basset, *The Tacoma News Tribune* editorial cartoonist Chris Britt, *Seattle Post Intelligencer (PI)* editorial cartoonist Dave Horsey, *Seattle PI* graphic artist and cartoonist Steve Greenberg, and the creator of *The Far Side*, Gary Larson, provide some thought provoking and entertaining environmental views for potential Streamkeepers to ponder.

The Washington State Department of Ecology, The Snohomish County Planning Department, and the Environmental Protection Agency provided financial support during the development of the 1991 edition of the book which was the prototype for this version. Special thanks to

Glen Crandal, Department of Ecology Grants Administrator, Tom Neiman, Snohomish County Planning Department Senior Planner, and Tom Wilson, EPA Chief of Water Planning Region X, for keeping the funds flowing when they were needed in that research and development stage.

Last, but certainly not least, a great deal of credit for this book belongs to past and current members of the AASF's Board of Directors and members of the AASF Environmental Education and Stream Enhancement Guild: Deberra Stecher, Tom Sasaki, Mike Chamblin, Russ Orrell, Dan Pebbles, Wolf Bauer, Kathy Sider, Grant Woodfield, Don Bayes, Bill Hershberger, Ernie Brannon, Bruce Bochte, Rick Garcia, Carl Shipley, Beverly Spillum, Elsie Sorgenfrie, Dick Walsh, and Terry Williams.

These folks not only helped raise funds to develop and support the Streamkeeper training program, but provided direction to transform this book from just another water quality monitoring textbook, into a publication that enables readers to take the actions necessary to protect and restore our waterways.

Contributing Artists

Artists' self portraits used with permission. All rights reserved.

PART ONE

Watersheds

CHAPTER ONE

Understanding Watersheds

A stream is only as healthy as its surrounding watershed. The amount of water carried by a stream, the shape of its channel, the chemical composition of its water, and its diverse life are determined by its watershed and what happens there. To fully understand your stream, you must look beyond its channels and learn about the land that surrounds it. Watersheds are fascinating.

INTRODUCTION TO WATERSHEDS

Each of us lives in a **watershed**, an area surrounding and "shedding" water into a stream, a river, a lake, or wetland. A watershed is also called a **catchment area** or a **drainage basin.** The system of streams which transports water, sediment and other materials from a watershed is called a **drainage network** or **drainage system**.

Watersheds and their drainage networks are interconnected land-water systems. A watershed catches water which falls to the earth as precipitation; a drainage system channels the water and substances it carries to a common **outlet**. The outlet of a watershed, or drainage system, is the **mouth** of the main stream or river. Its mouth may be where it flows into another river or stream, or the place where it empties into a lake, estuary, or ocean.

The Interconnected System
Although a river and its hillside do not resemble each other at first sight, they are pieces of the same puzzle. When this is appreciated, one may fairly extend the "river" all over the basin and up to its very divides. Ordinarily treated, river systems are like the veins of a leaf; broadly viewed, they are the entire leaf.

Like streams and rivers, watersheds vary in size. A watershed area can be very small, perhaps only a few square kilometers such as one that drains a small stream in your neighborhood. Or it can be very large, like the 668,220 square kilometer Columbia River watershed, or the even larger Mississippi River watershed.

Tributaries are smaller feeder streams or rivers which empty into a larger **mainstem river**. The watersheds of tributaries are referred to as **sub-watersheds** or **sub-basins**.

Thus large watersheds draining into rivers such as the Columbia or Mississippi may be made up of many smaller sub-watersheds. The smaller sub-watersheds may, in turn, be made up of even smaller watersheds that are fed by smaller tributaries. Any given watershed may be part of a larger watershed and itself be made up of smaller watersheds.

The Triple Divide
The headwaters of three of North America's largest rivers come together at the Triple Divide in Glacier National Park. This rocky mountain ridge forms the boundaries of three large watersheds which drain into three of the Earth's oceans. From the Triple Divide, the Saskatchewan and its tributaries flow northward into the Hudson River and to the Arctic Ocean. Tributary streams of the Mississippi and Missouri rivers flow southward and eventually the Mississippi empties into the Gulf of Mexico. Flowing westward, the tributaries of the Columbia River head toward the Pacific Ocean.

figure 1.1

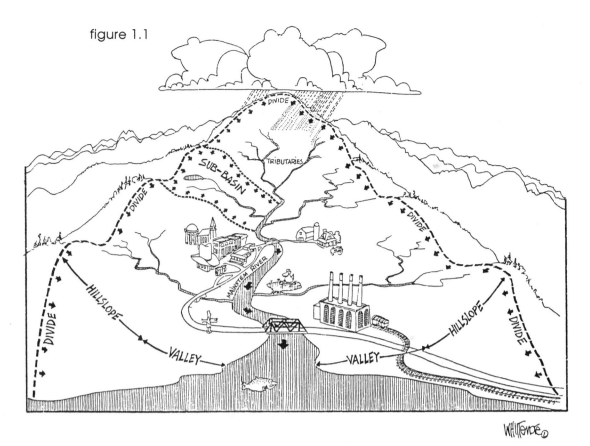

Topography, the three-dimensional shape of the land surface, determines the boundaries of all watersheds. The topographic boundaries of a watershed are called **divides.** They are the highest points surrounding a stream or river. If a drop of water falls on one side of a divide, it eventually drains into that watershed's stream or river. If the drop lands on the other side of the divide, it drains into the stream or river of a neighboring watershed.

No watershed is exactly like another; each has a different size, shape and drainage pattern. Two factors responsible for the varying natural characteristics of watersheds are **geology** and **climate.** Geologic processes, including glaciers, volcanoes and plate tectonics, determine the rock formations and how they change over time. Climatic processes erode and shape the rock formations.

Weathering and erosional agents such as rain, snow, wind, glaciers, and variation in temperature create topographic variation in watersheds. Streams result from precipitation and are important agents of surface erosion; they carve rocks into upland areas called **hillslopes** and lowland areas called **valleys.**

Through the process of carving valleys, streams break large rock material down into smaller pieces. They transport fine materials, or **sediment** throughout their watersheds. In fast flowing streams, sediment adds erosional force to the water, cutting through valleys and scouring stream bottoms. In slow flowing streams, sediment is deposited, settling out to form sandbars, levees, and fertile flood plains.

Like water, sediment is crucial to life. Sediment and organic matter comprise **soils,** the skin of a watershed and interface between the living and nonliving parts. Soils have different textures, mineral content and water holding and transmitting properties. They play a key role in watersheds by determining which plants grow, how much water runs off the land and how susceptible the land is to erosion.

Vegetation is another key watershed component. Plant roots slow and absorb runoff, releasing the water slowly to groundwater and streams and back into the atmosphere. Vegetation also plays many other important roles, including providing nutrients and habitat for fish and wildlife. Each watershed, with a distinctive combination of soils and plant communities, supports a diversity of habitats that support a diversity of life.

THE HYDROLOGIC CYCLE

Watersheds are part of a global, solar powered, water circulation network. The **hydrologic cycle** transports water between earth's watersheds, atmosphere and oceans. The various parts of each individual watershed are linked by the cycle as well. Water and other substances are transported and exchanged between surface and subsurface parts and between living and nonliving parts.

Water Above the Surface

The major processes moving water around in the hydrologic cycle are evaporation, transpiration, condensation, and precipitation. **Evaporation** occurs when the sun's energy turns liquid water on the earth's surface into water vapor. Water vapor, which is simply water in its gaseous form, enters the atmosphere. Water also enters the atmosphere through transpiration, the process of the sun's energy "pulling" water from plants and releasing it into the atmosphere as vapor.

Both evaporation and transpiration involve the movement of water from the earth's surface to the atmosphere. Because they are difficult to measure separately, they are collectively called **evapotranspiration**. More than two thirds of the water that falls to earth returns to the atmosphere by evapotranspiration.

Condensation happens when water vapor in the atmosphere cools to form clouds full of liquid droplets or particles of ice. When these droplets or particles become heavy enough to fall to earth, **precipitation** occurs.

Water on the Surface

Through precipitation, water enters watersheds on the earth's surface as rain or snow. Some precipitation does not stay long on the earth's surface because it is intercepted by plants and returns to the atmosphere through evaporation. This process is known as **interception**. Some water that falls on vegetation does reach the ground, however, because it drips off foliage and branches or runs down stems and trunks before it heats up enough to evaporate.

If precipitation does get to the ground, it may remain on the surface or be absorbed by the soil. Water that remains on the surface and eventually runs off into streams or lakes becomes **overland flow** or **surface runoff**. Water that seeps into the soil does so by entering the spaces between soil particles. This process is known as **infiltration**.

figure 1.2 Hydrologic Cycle

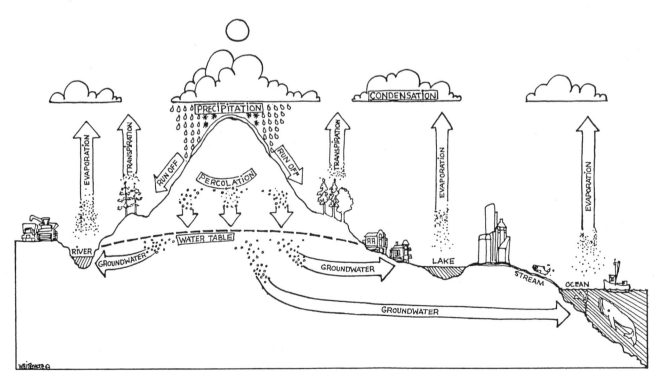

The Streamkeeper's Field Guide

Where on Earth is Water?

At any point in the hydrological cycle, more than 97% of the water is contained in the ocean, approximately 2.5% is stored on land, and .001% is contained in the atmosphere. While the amount of water in the atmosphere seems very low, it is a very active part. It is estimated that the atmosphere recycles its water every 8-10 days.

Of the 2.5% stored on land, approximately 79% is ice, 20% is in the ground, and 1% is on the surface. The ground and surface water not tied up in ice is the watershed water that your stream depends on.

Oceans	**97.4704%**
Land	**2.5286%**
Glaciers/Polar ice	2.0%
Groundwater	0.5%
Lakes	0.0175%
Soil moisture	0.011%
Rivers	0.0001%
Atmosphere	**0.001%**

Water Under the Surface

There are different paths the water can take after it infiltrates the soil. Some is absorbed by plant roots and eventually returned to the atmosphere by evapotranspiration. Water that is not absorbed by plants may move laterally into streams as **subsurface storm runoff**, or move downward into the **groundwater zone**. The groundwater zone is the area under the soil that is **saturated** with water - where the pore spaces between soil or rock particles are filled with water.

Aquifers are areas of groundwater that can store and transmit a significant amount of water. Aquifers are important sources of water in many places, and for some human populations, an aquifer may be the sole source of water. In the arid Yakima Basin of eastern Washington State, 75% of the drinking water comes from aquifers. In the wetter Snohomish Basin of western Wash-

ington, only 33% of the drinking water is provided by aquifers because surface water is more abundant.

Surface water and groundwater are intimately connected in watershed. Groundwater may eventually move into streams, lakes or wetlands through seeps and springs, sometimes taking hundreds or even thousands of years to reappear at the surface. The area where it comes to the surface is a **discharge area**. Discharge areas are often found in topographical low spots. In contrast, surface water in lakes, streams or wetlands may sink into the ground in a **recharge area** and then become groundwater.

The contribution of groundwater to surface water systems is known as **baseflow**. It can provide water during periods when there is little rain and surface runoff. In fact, during dry times, the lower elevation reaches of many streams consist only of groundwater discharge. Groundwater and surface water are part of the same watershed system. The only difference is where the water is at a particular time.

A CLOSER LOOK AT SURFACE WATERS

Streams, lakes, and wetlands represent less than .03% of the earth's terrestrial water. This amount seems small when compared to the portion of the earth's water that exists as ground water (0.5%, or 16 times that of surface water). It is even smaller when compared to the water trapped in glaciers and polar ice (2.0%, or 66 times that of surface water). Nevertheless, the water carried and stored in streams, lakes, and wetlands is very important. It is the water used and required by watershed life; it is the medium in which the crucial chemicals of life are cycled and exchanged in photosynthesis and respiration.

Lakes, streams and wetlands are part of the same watershed network; however, they represent very different physical, chemical, and biological conditions. One major factor which shapes stream, lake and wetland ecosystems is the nature of the **flow regime**, or the volume and velocity of water moving through a system.

Wetlands play a number of roles, including:

- controlling floodwaters and sediments
- filtering pollution from stormwater
- recharging ground-water aquifers
- discharging water to streams in dry seasons
- providing fish and wild-life habitat

Lakes, Ponds, and Wetlands

Lakes, ponds, and wetlands, which consist of relatively slow moving water, are called **lentic systems.** They have a longer **retention time** than streams and rivers. Lakes, ponds, and wetlands are usually more prone to pollution impacts than streams and rivers where water moves faster and pollutants can be flushed out more quickly.

Lentic systems are classified primarily by their size and depth. Lakes are usually more than six feet deep or greater than 20 acres in size; ponds are usually shallower and smaller. Lakes and ponds are types of wetlands, but the term wetland usually refers to systems such as **marshes, bogs** and **swamps.** These are transitional areas between terrestrial and aquatic environments. They may not be filled with water all year, but their soils are saturated at or below the surface.

As lakes and ponds age, they may gradually turn into wetlands, as their open water is filled with sediment and colonized by plants adapted to low oxygenated, wet soils. Eventually, if the process continues, wetlands may turn into upland meadows. This phenomenon, referred to as **lake succession,** does not always occur. Some systems may follow other trends, depending on their hydrology, nutrient inputs, sedimentation patterns, and many other factors.

Although they may not always have standing water in them, wetlands are as important to watersheds as lakes and ponds. Their water is stored in soils and plants and released over longer periods of time.

Wetlands are often adjacent to streams or lakes. Since they don't have a definite shoreline or stream bank, their boundaries are more difficult to define. Moist or **hydric** soils and water-tolerant plants are wetland indicators. By learning the various types of soils and plants endemic to wetlands you can determine their boundaries.

figure 1.3 Stream Ordering

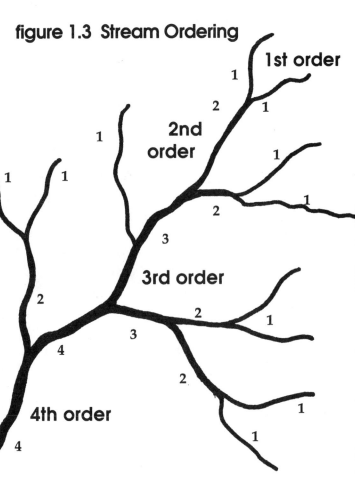

The Streamkeeper's Field Guide

Streams and Rivers

Streams and rivers which consist of relatively fast moving water are called **lotic systems.** Fast moving water influences a stream's physical and chemical characteristics, creating cool temperatures and high concentrations of oxygen. Organisms that are adapted to these conditions live in streams.

Streams are also classified by size. In any particular watershed, the smallest streams that have year round water and no tributaries are **first order** streams. When two first order streams come together they form a **second order** stream. Further along the course, a second order stream may join another second order stream to form a **third order** stream, and so on.

Note that when a first order stream joins a second order stream, the resulting stream remains a second order stream. A third order stream is only formed if two second order streams come together. A fourth order stream is formed when two third order streams flow together, and so on. . . you get the picture.

Streams also can be thought of as progressing from youth to old age. The **headwaters** where a stream originates are generally at a young stage. The lower reaches near the river mouth, where it empties into another water body, are usually mature. If you follow a river from its birth to maturity, you'll discover that it changes physically, with different habitats from top to bottom, and biologically, with different assemblages of animals. These changes occur along a gradual continuum.

The following section entitled "The River Continuum - An Imaginary Stream Tour," will lead you through a river system from headwaters to mouth. Though there is really no such thing as a typical river system, the river continuum description that follows provides a general model that many rivers exemplify.

THE RIVER CONTINUUM - AN IMAGINARY STREAM TOUR

Imagine you are flying above a rocky and forested region on the west or east side of the Rocky Mountains, near the crest. Like an eagle, you have the capability to focus at various **spatial scales,** zoom in for a closer look at the smaller things, and also to zoom out and look at the bigger picture.

Upper Reaches

You see several small headwater streams originating from a variety of sources---snow melt, surface runoff, a lake outlet, or ground water which surfaces as springs. Some may be **ephemeral,** flowing only during periods of intense rainfall or high ground water. Some are **intermittent,** flowing during the wet seasons of the year. Others are **perennial,** having year-round flow.

These headwater streams are classified as first order and second order streams. You zoom into one of these small, young streams. It is no more than a few feet wide. It has a fairly steep **gradient,** or vertical drop over a set distance. Various logs and boulders give it a stair-step appearance. Water cascades over them to form deep **plunge pools** or **step pools.** Cobble and gravel fill in around the boulders to cover the stream bottom.

The stream valley that contains the channel is narrow and "V" - shaped. The **riparian zone,** the area of vegetation adjacent to the stream, almost completely covers the stream with its **canopy.** This zone is fairly narrow and the vegetation is composed of some herbs, shrubs and trees; a few **deciduous** trees that have broad leaves and **conifers** that have needles.

Very little sun gets to the stream. The plant material that enters the stream in this headwater area comes from outside the stream. It consists of leaves, needles, and woody stems, branches and tree trunks known as **organic debris** and **woody debris.** This material supports the bottom of the food chain, providing food and habitat for the small organisms that are the **primary consumers** of the stream.

Middle Reaches

Downstream, in the middle reaches of the river system, some tributaries have entered the stream and added to the flow. It is bigger and deeper, now a third or fourth order stream. It

reflects an older system; its channel has widened into a "U" - shape and has eroded its stream banks laterally in some places. You can detect a **flood plain,** a lateral flat area along the stream banks. The stream periodically overflows onto this area, slows in flow, and dumps its load of sediment. The stream has begun to flow across its flood plain in curves or **meanders.**

The gradient of the stream has decreased. There is no longer a stair-step appearance. The stream still has logs across it, but they are farther apart and are usually accompanied by deeper areas of water called **pools.** In between the pools are shallower areas of faster moving water called **riffles,** where a few rocks break the surface. Pools and riffles alternately make up the stream in this area. The bottom substrate is composed of mostly gravel and cobble.

The riparian zone is a little wider; its canopy no longer reaches all the way across the stream. Organic debris still falls into the stream from the riparian zone to feed the primary consumers; however, some sunlight gets to the stream, allowing photosynthetic algae to grow and become part of the food base. The composition of the food base changes and nurtures a slightly different community of organisms than in the upper reaches.

Lower Reaches

Progressing downstream towards the stream's mouth, more tributaries have entered and added more flow to create a **mainstem river.** The wider, deeper channel reflects an older, mature stage. The river flows in big, arcing meanders through a flat flood plain and broad valley.

The bottom substrate of the river may consist of sand, gravel and mud. As the river flows toward its mouth, the load of sediment grows beyond the river's capacity to carry it. The sediments and debris form a **delta** as they are deposited. The river is forced to split and follow different paths, giving the channel a braided appearance.

Side channels, sloughs, and wetlands are interspersed throughout. The riparian zone is broad and complex, with different kinds of grasses, shrubs and trees. It only covers the sides of the river; most of the water is unshaded. The community

figure 1.4

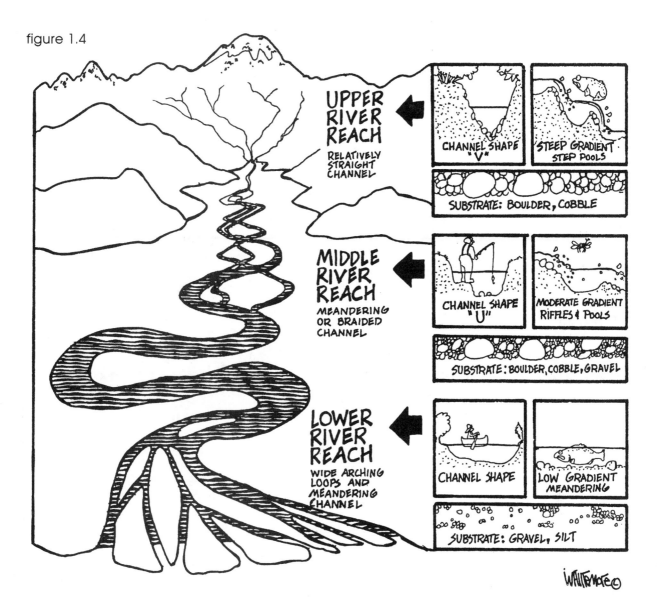

UPPER RIVER REACH
RELATIVELY STRAIGHT CHANNEL

CHANNEL SHAPE "V"

STEEP GRADIENT STEP POOLS

SUBSTRATE: BOULDER, COBBLE

MIDDLE RIVER REACH
MEANDERING OR BRAIDED CHANNEL

CHANNEL SHAPE "U"

MODERATE GRADIENT RIFFLES & POOLS

SUBSTRATE: BOULDER, COBBLE, GRAVEL

LOWER RIVER REACH
WIDE ARCHING LOOPS AND MEANDERING CHANNEL

CHANNEL SHAPE

LOW GRADIENT MEANDERING

SUBSTRATE: GRAVEL, SILT

WHITEMORE©

of small aquatic organisms has changed again. Although the water is unshaded, the turbidity from suspended sediments prevents sunlight from reaching the bottom. Fine particles replace organic debris and algae as the food source for primary consumers.

At its **mouth**, the river empties into another body of water and carries its remaining load of sediment, debris and other substances with it. The water body can be an estuary, a lake, or a larger river system.

HUMAN USE OF WATERSHEDS

Everyone is dependent on the resources watersheds provide. As human populations increase, the demand for water, food, housing, transportation, power, irrigation, and recreation grows. Human uses of land and water resources affect the ecological system in watersheds by changing the quantity and quality of water as well as by altering natural habitats. Some of these changes have brought improvements, while others have brought degradation.

Changes in the Hydrologic Cycle

A watershed is one interconnected land/water system. An action in one area of a watershed, even far up into its headwaters, can affect what happens in distant areas downstream. Any changes we make to the land surface or to the vegetation also changes some aspect of the hydrologic cycle.

Constructing **impervious surfaces** (streets, parking lots, rooftops, etc.) in a watershed affects

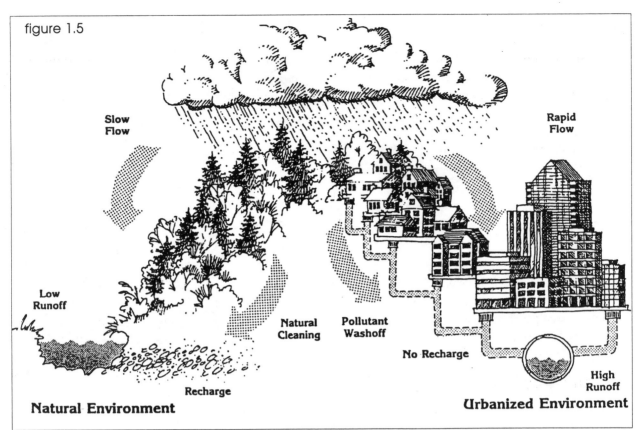

figure 1.5

Slow Flow

Rapid Flow

Low Runoff

Natural Cleaning

Pollutant Washoff

No Recharge

High Runoff

Recharge

Natural Environment

Urbanized Environment

drawing courtesy City of Bellevue Utilities Department

the hydrologic cycle in a watershed dramatically. Vegetation and wetlands act like sponges to slow and absorb water during wet times of the year. When we replace vegetation and wetlands with impervious surfaces, less water infiltrates into the ground and more water flows directly into streams via ditches and storm drainage pipes. The increased runoff may result in a variety of problems, including flooding, streambank erosion, sedimentation, down-cutting of streambeds, and pollution.

Problems caused by impervious surfaces are not only limited to the wet times of the year. Because infiltration is impeded, there is less recharge of groundwater. Thus, less water is discharged back to the surface during the dry times of the year, reducing base flow. Many streams simply dry up during periods of low rainfall.

In short, impervious surfaces reduce the **water storage capacity** of a watershed. When you can't store water, you end up having too much when it rains and not enough when it doesn't rain.

Humans also affect the hydrologic balance of a watershed when we divert water for domestic, agricultural, or energy needs. Dams provide

an obvious example of how the flow of rivers and streams is affected by human activity.

Although less obvious, pumping groundwater from wells also affects the hydrology of a watershed. Excessive groundwater use may lower the water table and reduce the amount of water discharged to surface water bodies.

The Streamkeeper's Field Guide

Water Pollution

While human uses of land and water have altered the quantity and timing of water cycling through watersheds, they have also impacted the quality of water resources. Increased runoff and water use may lead to increased water quality problems. Runoff carries sediment, nutrients, and other pollutants into surface water bodies. Increased water use results in less water available to dilute the concentration of pollutants.

Pollutants come in many forms, including: man-made chemicals such as pesticides and herbicides; heavy metals or radionuclides from mining activities; excessive additions of naturally occurring substances such as sediment and nutrients. Water pollution can result from any activity which changes the flow characteristics of streams or alters stream temperatures.

Point and Nonpoint Sources

Pollutants may be divided into two types: point and nonpoint, depending on their source.

Point source pollutants can be tracked to a specific source. Usually a point source is a pipe that discharges effluent into a receiving water body. Examples of point sources include:

- waste water discharges from industrial plants
- municipal sewage treatment plants
- urban storm sewer discharges
- thermal discharges from power plants, etc.
- animal feeding operations
- boat wastes

In older cities, stormwater and sewage are sometimes carried in combined sewer systems. During storms, combined sewage and stormwater can exceed the capacity of the treatment plants, overflowing directly into the receiving water body. Combined sewage overflows are composed of everything that goes down drains and runs off streets, from untreated sewage to toxic chemicals and heavy metals. The quantities can be staggering. For example, according to Seattle METRO, in 1994 the greater Seattle area discharged 2.4 billion gallons of untreated wastewater into Puget Sound, Lake Washington and the Duwamish River.

While the thought of water pollution brings to mind the image of chemicals oozing or sewage pouring out of pipes, not all forms of water pollution are point sources. **Nonpoint sources** are pollutants that arise from a number of sources rather than just one. They are typically generated by human activities all over a watershed and carried to streams, rivers, lakes, and wetlands by surface runoff. Nonpoint pollution is insidious and difficult to deal with because it is usually less obvious and more widespread.

Clearing land for residential, commercial and industrial use or agricultural and timber production increases runoff and nonpoint pollution. Often when it rains, farms, logging roads, parking lots and lawns send pesticides, fertilizers, oil and grease, hazardous metals and oxygen-robbing nutrients and sediments into the streams. During larger storms, runoff can be catastrophic, washing large quantities of sediment and nutrients into streams, along with toxic by-products of our daily activities.

Table 1 lists some common nonpoint pollutants, their diffuse sources, and some effects on water quality. Note that urban runoff is listed as a source for almost all the pollutants in the table. Our individual habits and actions contribute to the nonpoint pollution in urban runoff. Our lawns and gardens contribute nutrients, sediment, and pesticides; pet waste and septic tanks contribute nutrients and fecal bacteria; automobiles contribute petroleum, metals, and other toxic residues; and various cleaning solvents, paints, and other household toxic substances also end up in urban runoff.

Despite the large concern over toxic substances, the leading nonpoint source pollutants in watersheds are sediment, nutrients and fecal bacteria. It is the excess of these naturally occurring substances which causes many of the water quality problems in our watersheds.

GLASS   METAL   PAPER   INCRIMINATING EVIDENCE

TABLE 1 SOME NONPOINT SOURCE POLLUTANTS, THEIR SOURCES, AND WATER QUALITY IMPACTS

POLLUTANT	SOURCES	WATER QUALITY AND RELATED IMPACTS
sediment	agriculture - crops & grazing forestry urban runoff construction mining	• decreases water clarity and light transmission through water, which: - causes a decrease in aquatic plant production - obscures sources of food, habitats, refuges, and nesting sites of fish - interferes with fish behaviors which rely on sight, such as mating activities • adversely affects respiration of fish by clogging gills • fills gravel spaces in stream bottoms, smothering fish eggs and juveniles • inhibits feeding and respiration of macroinvertebrates, an important component of fish diets • decreases dissolved oxygen concentration • acts as a substrate for organic pollutants, including pesticides • decreases recreational, commercial and aesthetic values of streams • decreases quality of drinking water
pesticides **herbicides**	agriculture forestry urban runoff	• kill aquatic organisms that are not targets • adversely affect reproduction, growth, respiration, and development in aquatic organisms • reduce food supply and destroy habitat of aquatic species • accumulate in tissues of plants, macroinvertebrates and fish • some are carcinogenic (cause cancer), mutagenic (induce changes in genetic material-(DNA), and/or teratogenic (cause birth defects) • create health hazards for humans consuming contaminated fish or drinking water lower organisms' resistance and increase susceptibility to diseases and environmental stress • decreases photosynthesis in aquatic plants • reduces recreational and commercial activities
polychlorinated biphenyls (PCBs)	urban runoff landfills	• accumulate in tissues of plants, microinvertebrates and fish • toxic to aquatic life • adhere to sediments; persist in environment longer than most chlorinated compounds
polycyclic aromatic hydrocarbons (PAHs)	urban runoff	• accumulate in tissues of plants, macroinvertebrates and fish • when digested, create substances which are carcinogenic (cancer-causing) • toxic to aquatic life • toxicity is affected by salinity; estuaries with low salinities may be the most biologically sensitive
petroleum hydrocarbons	urban runoff	• water soluble components can be toxic to aquatic life • portions may adhere to organic matter and be deposited in sediment • may adversely affect biological functions

POLLUTANT	SOURCES	WATER QUALITY AND RELATED IMPACTS
pathogens & fecal bacteria	agriculture forestry urban runoff	• create human health hazard • increase costs of treating drinking water reduce recreational value

nutrients (phosphorus, nitrogen)	agriculture
	
	forestry
	
	urban runoff
	
	construction
		• overstimulate growth of algae and aquatic plants, which later, through their decay, cause:
		  -reduced oxygen levels that adversely affects fish and other aquatic organisms
		  -turbid conditions that eliminate habitat and food sources for aquatic organisms
		  -reduced recreational opportunities
		  -reduced water quality and increased costs of treatment
		  -a decline in sensitive fish species and an overabundance of nutrient-tolerant fish species, decreasing overall diversity of the fish community
		  -premature aging of streams, lakes and estuaries
		• high concentrations of nitrates can cause health problems in infants

metals	urban runoff industrial runoff mining automobile use	• adversely affect reproduction rates and life spans of aquatic organisms • adversely disrupt food chain in aquatic environments • accumulate in bottom sediments, posing risks to bottom feeding organisms • accumulate in tissues of plants, macroinvertebrates, and fish • reduce water quality
sulfates	mining industrial runoff	• lower pH (increase acidity) in streams which stresses aquatic life and leaches toxic metals out of sediment and rocks. • high acidity and concentrations of heavy metals can be fatal to aquatic organisms, may eliminate entire aquatic communities
radionuclides	mining and ore processing nuclear powerplant fuel and wastes commercial/ industry	• release radioactive substances into streams • some are toxic, carcinogenic (cancer causing) and mutagenic (induce change in genetic material-DNA) • some break down into "daughter" products, such as radium and lead, which are toxic and carcinogenic to aquatic organisms • some persist in the environment for thousands of years and continue to emit radiation • accumulate in tissues, bones and organs where they can continue to emit radiation
salts	agriculture mining urban runoff	• eliminate salt intolerant species, decreasing diversity • can fluctuate in concentration, adversely affecting both tolerant and intolerant species • impact stream habitats and plants which are food sources for macroinvertebrates • reduce crop yield • decrease quality of drinking water • reduce recreation values through high salinity levels and high evaporation rates

Cumulative Effects

We have discussed how day-to-day activities contribute to nonpoint source pollution. The impacts of human activities depend on the type of activity, its location, duration, and intensity. The impact also depends on the nature of the watershed, its topography, soils, vegetation, climate and drainage system. Each type of activity alters the watershed and contributes a set of pollutants to streams. Though individual contributions may be small, their combined effects can be significant.

The nature of the watershed will determine what types of human activities occur. Some watersheds may have vast mineral resources which are mined to generate power or make metal alloys. Others may have rich forest resources which are harvested to construct homes and to manufacture paper. Others may have recreational, biological and aesthetic values which influence human activities.

Most watersheds are affected by more than one type of human activity. Timber harvesting may predominate in forested headwater regions, while farming and urban activities may predominate in the flatter, lower areas near the mouth of a stream or river. Some activities contribute the same pollutants. For example, sediment is a by-product of timber harvesting,

Editor's note: Even in industrialized communities sewage treatment is often inadequate. In much of the world there is virtually none...

farming, and house construction. Pollutant impacts should be evaluated as a cumulative effect of multiple influences.

Pollution Regulation

In the U.S., the federal **Clean Water Act** directs the Environmental Protection Agency to set nationwide criteria and policies regarding **water quality**. All states are directed to develop water quality standards and policies that are equal to or more stringent than federal criteria. State standards set up different water quality classifications such as excellent, good, fair, etc. that they assign to their surface water bodies.

How a state classifies a particular water body depends on its desired uses, such as domestic water supply, fish habitat, contact recreation, etc. A particular water body may not actually be in the condition its classification demands because classifications are based on *desired*, not actual, uses.

State environmental agencies regulate **point sources** of pollution by requiring permits to discharge wastes into surface waters. The amount of pollution the operator is allowed to release is stipulated in the permit and is determined by the water quality classification of the water body. For each water quality classification, the state standards set minimum allowable levels or specific ranges for various pollutants, such as dissolved oxygen, fecal bacteria, nitrates, etc.

Usually operators that violate the stipulations in their permit are subject to a fine. Unfortunately, it's often more cost effective for violators to pay the fine than to retrofit their operation to meet water quality standards. Nevertheless, the U.S. has made significant strides over the last few decades in protecting water quality from point sources of pollution.

We've now realized that **nonpoint sources** of pollution have become the significant polluters of our watersheds. Controlling nonpoint pollution has proven to be more complex than dealing with point source pollution because it comes to our waters in the form of runoff that picks up sediment and pollutants from all over a watershed. This makes it difficult to point a finger at individual polluters.

In recent years, federal regulations have been developed in the U.S. to curb pollution problems arising from stormwater runoff. The **National Pollutant Discharge Elimination System Program (NPDES)** requires industries to develop stormwater pollution prevention plans. The program also requires municipalities with populations greater than 100,000 to implement best management practices to reduce stormwater pollution to meet water quality standards. However, there are many nonpoint pollution problems in areas of smaller populations. We still have a long way to go in meeting the challenges brought about by pollution-bearing stormwater.

Up until the more recent focus on nonpoint pollution in the U.S., the major government focus was on point source, or known sources of pollutant discharges, such as polluted water flowing from pipes. Government water quality standards have been based on chemical composition rather than ecosystem requirements. That is now changing. The U.S. Environmental Protection Agency has adapted a more holistic approach that considers physical and biological aspects of our nation's waters. While established water quality standards do not include ecological criteria such as aquatic organisms and habitat features, more emphasis is now being given to these factors.

Every Day Oil Spills

The Washington State Department of Ecology estimates that more than 4.5 million gallons of used motor oil is dumped into Washington's environment every year (enough to fill a medium sized tanker). Imagine the amount of oil spilled in watersheds throughout the United States and the cumulative impact on our water resources! It only takes one <u>pint</u> of oil to cause an oil slick one acre in size on your local lake. One <u>quart</u> of oil will foul the taste of 250,000 gallons of drinking water. Think of what it can do to aquatic life. (Of course, recycling used motor oil is the obvious alternative to dumping it.)

Maybe congress should ask us stream dwellers what the water quality standards should be...

## CHAPTER TWO

Watershed Investigations

Starting down the path to healthier watersheds and streams can begin with investigating your own neighborhood watershed and streams. Learning about a stream and its watershed can be a great adventure! Check out a stream that flows through your neighborhood, a stream with a favorite fishing hole, a stream near your school or workplace. Pick out that stream that is "near and dear" to you.

Streamkeepers begin by investigating their watersheds before looking more closely at their streams. An **inventory** is simply an assessment. It helps determine and document the physical, chemical, and biological , as well as the social and political characteristics of your watershed. By inventorying a watershed, it becomes possible to get to know its many natural and cultural resources, its uses by humans and wildlife, and the health of its waters.

As you have seen, everything that happens in the watershed affects the stream. A watershed inventory will give you some clues as to what's happening in the land that surrounds and drains into your stream and how the various activities might be affecting it. It will help you better design your stream monitoring objectives.

Investigating your watershed is essentially a three part process:

1. First, identify the boundaries of your watershed and make a working map.
2. Then gather all the available information you can find about your watershed.
3. Next, get out and conduct a field inventory of your watershed.

Because our world is constantly changing, you'll find that you can be most effective if you repeat steps two and three on a yearly basis. This way you can keep abreast of changes that occur in your watershed and adapt your stream monitoring activities accordingly.

Before you start your inventory, determine the area where you will focus your attention. The actual watershed area depends on the stream and outlet point chosen. Your watershed may be a sub-basin to a larger river basin. In this case, your outlet point will be the place where your stream flows into the larger river (the **confluence** of your stream with the river). In other cases, your outlet point may be where your stream flows into a lake or body of salt water.

If you choose to work on a larger stream or river system, your "outlet" point may not be a real outlet, but simply a point along the waterway, and your focus will be the watershed upstream from that point. If you focus on a relatively large stream or river system, it may be difficult for one individual or small group to cover the entire watershed. In this case, it is best if you combine your efforts with other groups or individuals in the upper and lower reaches of that system's watershed so that the whole basin is covered, from the headwaters to the mouth of the river, (see figure 2.1).

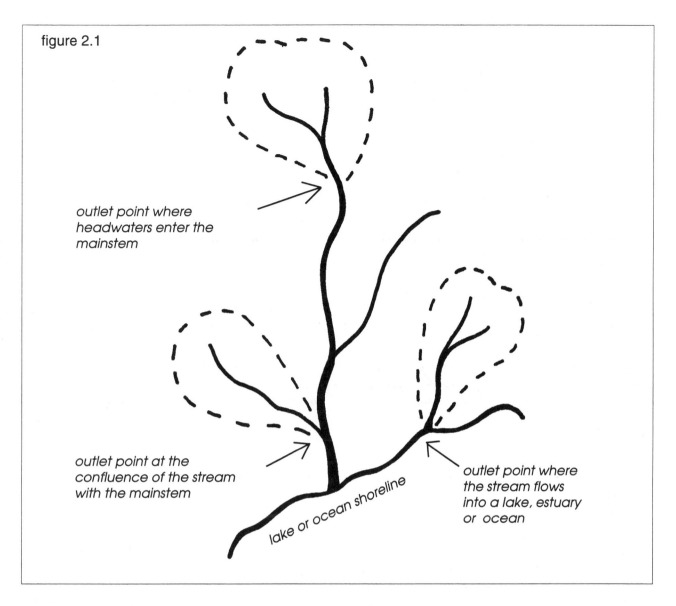

figure 2.1

outlet point where
headwaters enter the
mainstem

outlet point at the
confluence of the stream
with the mainstem

lake or ocean shoreline

outlet point where
the stream flows
into a lake, estuary
or ocean

WATERSHED MAPS

There are many choices when it comes to maps that might be useful to you. The most important tool to start with is the "topo" map. Topographic maps can be used to determine watershed boundaries.

Topographic Maps

Topographic maps show the shape and elevation of land forms with **contour lines**. All points along any one contour line are at the same elevation. Read, "Understanding a Contour Line" on the next page to get a start on how to understand topographic maps.

In the U.S. we recommend you use topographic maps published by the United States Geographical Survey (USGS). They are frequently used and referenced. A road map will not work. USGS "topo" maps portray a specific area of land or **quadrangle,** a four-sided region bounded by lines of latitude and longitude. Degrees of latitude and longitude are shown in the corners and pinpoint the area's location on the globe. (See Chapter 3 for more information on latitude and longitude.)

Different scales of topo maps cover different size quads. The most detailed maps from the USGS are in the **7.5 minute series**. A map from this series has a scale of 1:24,000 (one inch on the map equals 24,000 inches on the ground). The area of a 7.5 minute map is approximately 7 miles long and 7 miles wide.

Understanding A Contour Line

Maps are flat, but the areas they represent are filled with hills, valleys, mountains, and plains. Contour lines represent points of equal elevation. The lines allow mapmakers to show the "lay of the land" very clearly. The following exercise will help those of you who have a hard time seeing three dimensions on a flat map. It is taken from the 1990 Boy Scout Handbook, which also has great tips on other vital Streamkeeper mapping skills, like how to use a compass.

"To understand how contour lines work, make a fist with one hand. A fist has width, length, and height, just like the land. With a water soluable pen or magic marker, draw a level circle around your highest knuckle. Draw a second circle just below that one. Start a third line a little lower. Notice that to stay level, the pen may trace around another knuckle before the third circle is closed. Continue to draw circles, each one beneath the last. Lines will wander in and out of the 'valleys' between your fingers, over the 'broad slope' on the back of your hand and across the 'steep cliffs' above your thumb. After all the lines are drawn, spread your hand flat. Now it has only width and length, just like a map. But by looking at the contour lines you can still imagine the shape of your fist. Small circles show the tops of your knuckles. Lines that are close together indicate steep areas. Lines farther apart show the more gentle slopes of your hand. The contour lines of a map represent terrain in the same way. Small circles are the tops of hills. Where the lines are far apart, the ground slopes gently. Where they are close together, a hillside is steep."

Take this exercise a step further. Repeat the procedure with your other hand. Putting your two fists side by side demonstrates the boundaries of a watershed. You will have two high areas or ridges, with a valley in between. The bottom of the valley is where you would expect to find a stream.

Understanding how the contour lines represent valleys and ridges may be the most challenging thing about reading topo maps. Contour lines that represent a valley or depression usually are V- or U-shaped, with the tips of the V's pointing toward higher elevations. Lines that show a ridge are also shaped like V's or U's, but the tips point toward lower elevations. Water flows through the valleys perpendicular to contour lines.

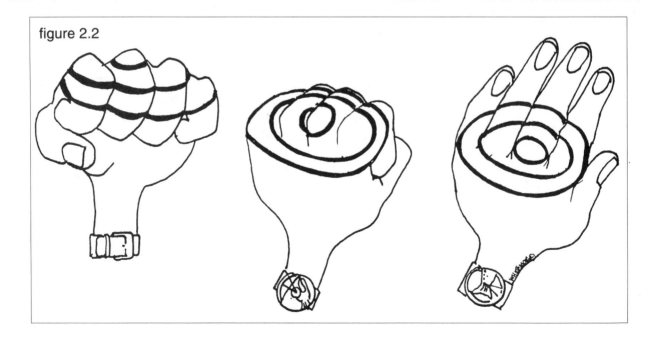

figure 2.2

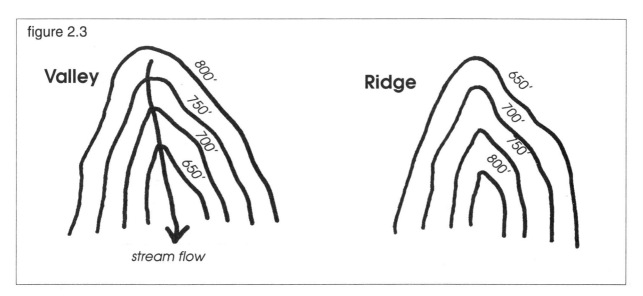

figure 2.3

Valley

800'
750'
700'
650'

stream flow

Ridge

650'
700'
750'
800'

USGS maps convey some basic information on land surface features, such as elevation, location of water bodies, vegetation, and some land uses. Colors indicate specific features: blue for water features, brown for contour lines, green for vegetation, red for roads and surveys, etc.

Symbols also give key information on topo maps. Some symbols look like the object they represent, e.g. a pick and ax represent a mine or quarry. One important symbol to look for is the one used to denote the location of gauging stations, a circle sectioned off into four quarters, two of them black and two of them white. These are stations where the USGS collects continuous information on the volume of water that is flowing in the stream or river. If a gauging station is located in your watershed, the data it generates will be quite useful if you decide to monitor the flow of water in your stream (see the section on flow in Chapter 6).

Many reports use features from USGS maps, such as **township** and **range lines** to describe locations in watersheds. These are numbered vertical and horizontal lines much like latitude and longitude. (See Chapter 3 for more information on Latitude and Longitude).

Topo maps can be purchased from local map stores, sporting goods stores and camping stores. If your watershed is large, you may need more than one quadrangle map. Most map stores will have an index of available maps for your state. They should also have a useful guide to the colors and symbols for your USGS map, entitled "Topographic Map Symbols."

You can also get topo maps from your local city, town, and county planning or public works departments. These local sources often have topo maps at much smaller scales that may provide you with greater detail on your watershed boundaries.

And don't forget your state resource management agencies. For example, Washington State's Department of Natural Resources publishes a catalog of maps and aerial photographs. You can find similar resources in most states. If topo maps are not available in your area, you can call 1-800-USA-MAPS for information on maps across the country.

WHO NEEDS A MAP! I KNOW THIS COUNTRY LIKE THE BACK OF MY HAND.

HE MAY HAVE TO EAT THOSE WORDS.

THE DONNER PARTY'S FAMOUS LAST WORDS

Sections, Townships, and Ranges

When you pay a visit or make a call to an agency requesting a map, they may ask you to locate the area you are interested in by referring to its section, township, and range numbers. You should become familiar with this "locator" information. Most USGS maps have a series of squares overlaying the geographic features on the map. These squares are referred to as townships, which are laid out in a grid format. Townships are 36 square miles in size.

Townships are identified by township and range numbers. Township numbers correspond to horizontal rows, while range numbers identify vertical columns. Thus township numbers appear along the sides of the map and range numbers appear along the top of the map. Township and range numbers will also have a reference to the compass, e.g., North (N), South (S), East (E), West (W).

A section is a one square mile area within a township. Each township's 36 square miles (or 36 sections), are shown in a grid pattern with a different number corresponding to each section. On larger scale maps, each section is generally numbered; however, the entire township may not be shown. If the section numbers appear to be out of sequence, you have "traveled" into another township. More often than not you will be using the larger scale maps which provide greater detail.

Obtaining the locator information for your stream is a fairly simple process. If sections are not shown on the maps you obtain, you can make up your own plastic overlay of a township. Draw 36 square sections on the overlay and number them like the graphic below. Be sure that your overlay fits the scale of your map so you can match it up directly with each township.

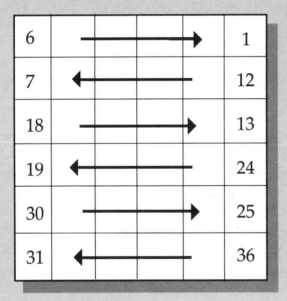

With a little practice you will become a map "wizard" and be able to locate and identify "anywhere USA" using this process. If you don't have the time to practice, at least you are now armed with a basic understanding of the process and will be able to ask for assistance in a knowledgeable manner.

Other Useful Maps

Aerial Photograph Maps usually are more recent than USGS maps. By comparing aerial photograph maps of different dates, you can see land use changes that occurred during the interim. They are also taken in larger scales, or in more detail than USGS maps. A common scale is 1:4800 (one inch equals 4800 inches, or 400 feet, on the ground).

Aerial maps will be especially helpful for getting around your watershed and obtaining land use information. They are available from private companies, state agencies and the federal government (the USGS), but you may find that your local planning departments are the best sources of aerial photographs.

For example, the Snohomish County Planning Department in Washington State has mapped the majority of the County's streams and wetlands onto 1:4800 scale aerial photographs. These maps are available to the public. They also will share **infrared** (distinguishes water, vegetation, and other features on the land with different colors) and **stereo**

photographs (black and white photographs put side by side that are viewed with a stereoscope to provide a three dimensional view). The Snohomish County Public Works Department has aerial photographs of the Snohomish and Stillaguamish River systems that date back to the 1930's which enable comparisons that provide great historical perspectives.

Some of the more sophisticated local governments even have aerial photographs overlaid with topographic contour lines that are at very large scales. You might find similar resources for your watershed.

Thomas Brothers maps are published in most states. These maps show detailed locations of roads in most watersheds. Most real estate agents and land developers use these maps to guide themselves around your watershed. You might consider getting these valuable tools for your use as well. They are available at many local book stores. This company also might be found in the yellow pages.

Property Ownership Maps are often available from your local city or county assessor's office. Typically these maps are very large scale (1 inch = 200 feet) and identify all property lines. The names of the landowners are also available at the assessor's office. It is public information.

Zoning Maps tell you what type of land use is allowed within your watershed. Locations of residential, commercial, and industrial development are usually shown along with areas set aside for timber harvest, agriculture, and mineral extraction. These maps, which are usually at the same scale as the assessor's maps, are generally available from city, town, and county planning departments.

Comprehensive Plan Maps show you what type of development you can expect to occur in your watershed during the next five to ten years. These maps, which are often printed in a variety of scales, are generally available from your local government planning department.

Fish and Wildlife Maps are often available from federal and state resource management agencies. You can locate current and historical distributions of fish and wildlife on these maps. Don't overlook your electric utilities as sources of fish and wildlife maps. For example, the Bonneville Power Administration, the utility that operates Columbia River dams, has a wide variety of maps available on all of the rivers and streams in Washington, Oregon, Idaho, and Montana. The Tennessee Valley Administration has similar maps in its service area . . . and so on.

Aeronautical Charts are available and are generally presented in a 1:500,000 scale. However, larger scale maps are available for heavily populated areas. Topographic features are highlighted in color. While these maps are not terribly useful for Streamkeeper field work, they do have a grid overlay in latitude and longitude that could prove valuable, and they can give you the big picture. These maps can be obtained at most local airports or at stores that cater to private pilots. You can get maps that cover up to a two-state area for as little as $5.

Demographic Distribution Maps show population centers and concentrations of people. These maps are usually broken out by **census tracts** and provide a variety of information including socio/economic factors that may be of

Map Scales

Maps come in a variety of scales. Scales express the ratio of the distance on the map to the actual distance on the ground. For example, a map with a scale of 1:1 would be life size. A map with a scale of 1:200 means that 1 inch on the map equals 200 inches on the ground. This is a larger scale map than one with a scale of 1:1000. A larger scale map shows a smaller area but more detail. To understand this intuitively, express the scale ratio as a fraction. Because 1/200 is a larger fraction than 1/1000, 1:200 is a larger scale than 1:1000.

interest to you. Local government planning departments and the Census Bureau at the federal government level can provide you with this kind of information.

Geographical Information Service Maps ("GIS" Maps) are maps generated from computer databases. They illustrate a variety of information, depending on what is contained in the databases. GIS maps may show watershed boundaries, land uses, aquifers, sources of pollution, fish populations, etc.

As you can see, there are a large number of tools that you may want to acquire before you launch a watershed inventory. You can also see that you will have to get to know some of the bureaucrats and agency representatives. You will find that these folks can become great allies for your cause, and if properly approached, will provide free sources for all of the maps and other tools that have been described here. Try to get as many maps and tools as you can. Each will contribute to a richer picture of your watershed.

CREATING YOUR OWN WATERSHED MAP

If you cannot locate a map illustrating the boundaries of your watershed, you can construct one using a USGS topographic map. Even if you can find a map with your watershed delineated, the following procedure is a useful exercise that will help you become more familiar with your watershed and its boundaries.

Determining the Boundaries of Your Watershed

1. Locate the outlet point of your watershed. It will be the lowest elevation in your watershed and in most cases will be the mouth of your stream.

2. Trace the stream from its mouth to its tributaries. Using a pencil, make marks along the stream and its tributaries every inch or so, dividing them into one inch sections.

3. At each mark, draw a line perpendicular to the stream or tributary, running out in both directions.

4. Follow each line out from the stream or tributary until you reach a maximum elevation. Mark all these high elevation spots with an "X."

5. Locate the beginning of each tributary, or the place where the stream's water originates. Extend a line out from each of these locations, in the direction opposite to the flow of water. Follow these lines until you reach a maximum elevation. Mark the high points with an "X."

6. Connect all the high points with a line, following ridges and crossing slopes at right angles to contour lines. The line resulting from "connecting the dots" will be the boundary of your watershed. Double check your boundaries to ensure accuracy, and then mark them with a pen or magic marker.

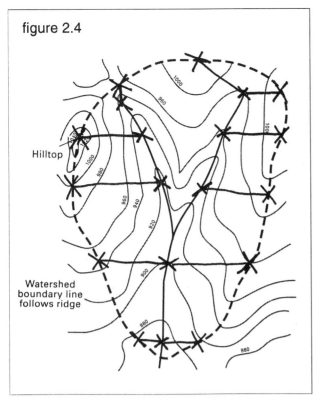

figure 2.4

Hilltop

Watershed
boundary line
follows ridge

Idealized Watershed Boundary. Adapted from *Method for the
Comparative Evaluation of Nontidal Wetlands in New Hampshire.*

Highlighting
Features on Your Map

Highlight your stream and its tributaries. You
will be able to see the pattern of the stream network. Count the number of tributaries and classify
the various streams into their orders.

Since you know the scale of the map, you can
estimate the length of each stream. Lay a piece of
string along the scale at the bottom of your map
and mark off the inches. Next, lay the string along
your stream and estimate its length. Note the locations where tributaries enter your main stream.

Pinpoint the other surface water features, such as
lakes and wetlands. Also trace the roads and high-
ways. How many roads cross your watershed? What
political boundaries, such as counties, cities, states,
and countries does the watershed cross? Locate other
features of interest, such as geological formations.

Other Points to Consider

It is a good idea to make some photocopies of
your watershed map. These copies will be working maps: you can label or mark the copies with
different features, activities and land uses or make

clear overlays to layer onto a base map. Remember
to make scale adjustments if you change the size of
the map when you photocopy.

You will want to include a copy of your map in
your data reports, take it out in the field when you
survey your watershed to ground proof some of the
map's features, and use it when you do a stream
survey. Attach a copy to Data Sheet 1, "Watershed
Inventory."

If you have adopted a tributary to a larger river,
you may want to obtain a map of that larger river
system's watershed. It's always good to know what
major river system ultimately carries the water in your
stream to the ocean, and to know how your stream
fits into that river system's watershed.

COLLECTING
WATERSHED INFORMATION

Now that you're familiar with your watershed's dimensions and resources, you're ready to
begin detailed investigative work. The information you collect about your watershed will come in
many different forms, from many different sources, and cover a variety of topics.

Data Sheet 1, "Watershed Inventory," contains
suggested topics and specific information to include in your watershed inventory, from physical
and biological information such as climate, geology, and vegetation, to social and political information such as the history, land uses, and
demography of your watershed. Use this data
sheet as a guide, but you need not limit yourself to
the topics it addresses.

The information you gather will provide you
with basic knowledge and a historical perspective
on your watershed. The information will also serve
as excellent background material for others want-
ing to get involved in a watershed project.

You may be amazed at the information that you
can "scrounge" up simply by making a few phone
calls and trips to local offices and libraries. These calls
and trips could lead to a few more, which could lead
to a few more . . . before you know it, you could be
drowning in information about your watershed. The
following are suggestions on where to look and who
to ask for information:

Physical and Biological Information

The nearest **National Weather Service office** can provide you with useful climate information. Climate data is important because it determines how much water enters your watershed from the atmosphere. Storm and flood records can give you clues about significant hydrologic events and patterns in your watershed.

The **U.S. Geological Survey's Water Resources Division (USGS)** has been collecting information on stream flow of large rivers for 50 years or more, in cooperation with state and local agencies. They also conduct studies of smaller streams for special reasons, such as urban runoff or floods from small drainages. They have one or more offices in every state. Check with the local or state office to see if they have information on your river or stream. They may be able to supply information on the magnitude of floods and information on baseflow and the frequency of low flow. They may also have information on water temperature and water chemistry.

Contact your local office of the **U.S. Natural Resources Conservation Service** (formerly the Soil Conservation Service) for soil maps of your watershed. In addition to information on soil types, they may also provide aerial photographs and information on vegetation, geologic features, water resources, land uses, and other topics.

A telephone call or two to your state's **fisheries and/or wildlife agency** should provide you with information on what "finny, furry and feathery" critters inhabit your watershed. **Electric utility companies and agencies** may also have extensive information on fish and wildlife populations, especially in regions utilizing hydroelectric power.

You can find out what changes in fish and wildlife populations have taken place over time and get a professional opinion on why these changes have occurred. Invariably, declines are due in part to habitat

loss, so be sure to ask your fisheries and/or wildlife biologists about critical habitats in your watershed. You'll also want to find out if your watershed is home to any endangered or threatened fish or wildlife species.

Your local city and county agencies may be able to provide physical and biological information about your watershed. A good deal of information on the items listed on Data Sheet 1 is available through your local planning department.

You'll also want to contact your **city or county public works and health departments** for information about water quality, water supply and domestic use, storm drainage systems, and sew-

age treatment systems. Some cities and counties have very active watershed related public involvement programs that can provide numerous resources for Streamkeeper groups.

Colleges and universities provide an excellent source of physcial and biological information about watersheds. You may want to contact professors and other university staff from several different departments to answer your questions about different aspects of the watershed. Check with departments such as water resources, natural resources, environmental studies, geology, forestry, agriculture, zoology, fisheries, etc.

Environmental, fishing, agricultural, and civic groups may provide useful information on the physical and biological aspects of watersheds. Information from these groups is often on a "grass-roots" level, and thus may be much more localized and current than information gathered from universities and government agencies.

These groups may also have information about social and political aspects of watersheds, such as local land use issues, historical information, and regulations pertaining to watershed and stream protection.

Historical Information

Visit a **library or local museum** to dig up some of the history of your watershed. Or, better yet, talk personally to the "old timers" and others who have spent their lives in the area. **Oral histories** can provide some terrific first hand accounts of "how things used to be." Historical information is invaluable in providing insight on how land use, vegetation, wetlands, and even the course of rivers and streams have changed over time in your watershed. Don't forget to tap into the information and resources available through **Native American (U.S.) and First Nations (Canada) elders,** if there is an Indian tribe within or close to your watershed.

Political and Regulatory Information

Find out what kind of federal, state and local government acts, laws, policies, regulations, ordinances, etc. are on the books that provide protection for aquatic systems and wildlife habitat. Dig deeper and find out how these rules are enforced or not . . . and why.

Because the business of regulations can be very complex, we recommend that you make a table that summarizes what you find and attach it to Data Sheet 1. The table could list different regulations by federal, state, county, and city, and include information such as: what agency implements and enforces the regulation; the purpose of the regulation; the area affected by the regulation; specific implications of the regulation to your stream and watershed; and problems that exist in enforcing the regulation.

In the U.S., the federal **Environmental Protection Agency** sets nationwide criteria and policies regarding water quality. States are directed to set and enforce water quality standards that are as strict or more strict than federal criteria. On the state level, water quality is usually regulated and enforced by state environmental agencies. These agencies are known by various names such as Department or Division of "Environmental Quality," "Environmental Management," "Ecology," etc. Contact your state environmental agency for more information on the regulation of water quality in your state.

States also manage water use. Water quantity issues are often treated separately from water quality issues, although the amount of water in a stream or river usually has a direct effect on the concentration of pollutants. And of course, the amount of water available is a crucial factor for aquatic life. State agencies may set **instream flow minimums** for critical times of the year. Contact your state environmental department or state water resources department for information about water use regulations.

When you contact your **state fish and wildlife agencies** for information about creatures inhabiting your watershed, also ask about how much authority these agencies have to regulate activities that affect fish and wildlife habitat in your watershed. In most cases, these agencies were originally set up to manage fishing and hunting. Now most play a role in protecting and restoring fish and wildlife populations and their habitats. They often follow guidelines and policies outlined by their federal parent agency, the **U.S. Fish & Wildlife Service**.

Many activities that affect watersheds are regulated by local political jurisdictions such as cities, towns, counties, districts, etc. Use your maps to determine what jurisdictions cover the area of your watershed. As water follows the natural boundaries of the land and not the artificial borders created by humans, watersheds can be complex politically, flowing from one city or county, into another.

Local governments may develop even stricter regulations than state governments or simply adopt what is dictated by the state. A trip to your

city or county planning department will shed light on local zoning ordinances, which tell you where commercial, industrial, and residential development is allowed. Ask for information on other ordinances that deal with aquatic resources and critical ecological areas. In addition, you can get copies of comprehensive land use plans. These are policy documents designed to guide how land use will occur in the future.

Something all effective Streamkeepers should know is the name, address, and telephone number of your local government elected officials. City,

town, and county council representatives are key resource managers in your watershed. They make vital land use decisions - adopting, amending, and sometimes removing laws and regulations that affect the integrity of streams and watersheds. Make a list of the elected officials who preside over your watershed and attach it to Data Sheet 1.

Land Ownership Information

Unless your watershed is in a remote area, there will be people living within its boundaries and probably alongside its rivers, streams and/or lakes. By visiting your **city or county assessor's office** you can find out who owns all the property in your watershed. Property owners are the *real* resource managers in a watershed.

If any public land holdings fall within your watershed's boundaries, contact the local, state, or federal agency which oversees that land. Federal agencies such as the **National Forest Service** or **Bureau of Land Management**, state agencies such as **Department of Natural Resources or Lands**, or local agencies such as **city or county parks departments**, usually have copious information on geol-

ogy, topography, soils, vegetation, water resources, fisheries and wildlife, and land uses.

You can see that watershed information comes from a wealth of sources. This is because there are so many facets to understanding a watershed.

In some cases, government agencies or private organizations develop **watershed management plans** and thus will compile a broad spectrum of information on a particular watershed. In most cases, however, you will find that no single entity collects all the information about a watershed, especially if the watershed crosses a number of political boundaries. City, county, state, and federal agencies collect similar data, but only for their piece of the watershed. One agency does not always know what the others know.

In addition, many smaller streams get lost in the shuffle of information. The U.S. Environmental Protection Agency has a **River Information System** covering rivers in each state in the country. The system lists information on many categories such as fish and wildlife populations, habitats, water quality, and recreational uses. However, only streams that show up on the 1:100,000 scale USGS maps are included in this resource database.

It is your task, as Streamkeeper, to find out what's out there and pool together as much information as you can about your watershed. It will require some patient investigative work and it will not be gathered all at once. Learning about your watershed will be an ongoing task.

It will be more fun and less work if you get other interested people to help you. Different members of your watershed group will have different interests and areas of expertise. They can collect various types of information. You could collect all the land use information from your county agencies. Your neighbor, an avid fisherman, could obtain fish population information from state agencies, and so on.

THE WATERSHED INVESTIGATION

FOLLOW THE PROPERTY LINES.

DEEP TROUT

As your bounty of information grows, it will be useful to find a central storage location such as a basement, library or a community center. If it is organized in one place, it can be used more easily by others who are interested.

Table 2 lists sources of watershed information, along with the kinds of information each source would most likely provide. It may be possible to obtain much of the information you will need either by telephone or writing to the source. Sometimes it becomes necessary to visit the information source to discuss what you need and look through available resources.

Keep in mind that any information produced by a public agency is public information. You have a right to access this information through the **Freedom of Information Act.** Some information is free while other resources may require a fee.

CONDUCTING A FIELD INVENTORY

By now you know within which major river system your watershed is located. You have made a map and have pulled together a lot of information from a variety of sources. Now it's time to get out in the watershed and look at it firsthand, to visually "ground truth" the information in all of the maps and reports you have researched. A field inventory is really the best way to get to know your watershed more intimately. Your most valuable tool will be your power of observation.

You may observe a seldom seen wildlife species or a rare plant. Perhaps you'll see a person dumping oil down a storm drain after an oil change or some cows in the middle of your stream. You may find that a tributary to your stream has changed its channel location since the last government map was made. Your observations will help you more fully complete data sheet 1, and will give you a lot of information to add to your watershed map.

Before you head out into the field, familiarize yourself with the section on page 34 entitled "Miss Mayfly's Guide to Watershed Etiquette, Safety and Liability Issues," a list of suggestions on how to least disturb your watershed and be safe and responsible during your inventory activities.

TRUST YOUR INSTINCT.

TABLE 2   WATERSHED INFORMATION

CATEGORY

SOURCE	CLIMATE	GEOLOGY	TOPOGRAPHY	SOILS	VEGETATION	GROUNDWATER	SURFACE WATER	WILDLIFE	FISHERIES	HISTORICAL	RECREATIONAL	LAND USE	DEMOGRAPHY	MINING	EDUCATION
COLLEGES/ UNIVERSITIES Depts of: Biology, Botany, Ecology, Entomology, Environmental Studies, Fisheries, Geology, Natural Resources, Wildlife, Zoology	*	*	*	*	*	*	*	*	*	*					
LOCAL AGENCIES **Libraries:** City, County	*	*	*	*	*	*	*	*	*	*	*	*	*	*	
Cities: Depts. of PublicWorks, Public Health, Planning				*	*	*	*	*	*	*	*	*	*		*
Counties: Depts. of Public Works, Planning, Public Health, Government Councils, County Extensions, Conservation Districts, River Basin Teams				*	*	*	*	*	*	*	*	*	*		
Indian Tribes: Fish and Wildlife Depts., Tribal Councils						*	*	*	*	*	*				
ENVIRONMENTAL GROUPS Check yellow pages under Environmental, Conservation & Ecological Organizations					*		*	*			*				*
REGIONAL River Basin Watershed Teams, Water Quality Authorities, Watershed Councils				*	*	*	*	*	*			*			*
STATE AGENCIES State Library	*	*	*	*	*	*	*	*	*	*	*	*	*	*	*
Dept. of Ecology						*	*					*		*	*
Dept. of Natural Resources, Lands, etc.			*		*		*		*	*		*	*		*
Dept. of Fisheries/ Wildlife						*	*	*	*		*				*
Dept. of Social and Health Services						*	*							*	
Office of Environmental Education															*
FEDERAL AGENCIES National Marine Fisheries								*	*						*
Environmental Protection Agency						*	*								*
Army Corps of Engineers		*	*	*	*	*	*					*			
Forest Service		*	*	*	*			*	*		*	*			*
Soil Conservation Service				*	*	*					*	*		*	
Fish and Wildlife Service					*			*	*		*				
Bureau of Land Management		*	*	*								*		*	
Geological Survey		*	*	*		*	*					*		*	*
National Weather Service	*														

FIELD PROCEDURE: WATERSHED INVENTORY

EQUIPMENT NEEDED

your senses: sight, hearing, smell, etc.
other people to help you
your watershed map
compass
Data Sheet 1, "Watershed Inventory"
clipboard & paper, preferably waterproof
pencils
binoculars (optional)
camera (optional)

WHAT?

The field inventory procedure is designed for you to gather more complete information on the items listed on Data Sheet 1. These items include: climate, geology, topography, soils, water resources, vegetation, wildlife, fish, historical aspects, land and water use, political/demographic factors, and water quality concerns.

Please note that this procedure is an overall watershed field inventory. Observations are meant to be general in nature, covering the entire watershed. You will be making more detailed observations about your stream during the stream reach survey, outlined in Chapter 4.

WHERE?

At this point you should have narrowed your investigation down to a sub-basin of a major river watershed. We recommend you focus your energy on a small watershed, because in this procedure, you will attempt to cover the entire watershed, not just the stream corridor.

WHEN?

After your initial inventory, we recommend you update your information on an annual basis. You'll find that a watershed field inventory is really an ongoing process. Every time you venture out into your watershed, you'll see new things that may be important.

HOW?

Getting Around Your Watershed

To conduct your field inventory, you may drive, walk or bike around your watershed, depending on the watershed size, roads, trails and weather. Though it is preferable to walk around your watershed for close, quiet observations, the area may be very large and some areas may not have trails. You might want to drive around parts of it and walk around others. If you have the resources or connections, find a private pilot who can take you on an aerial tour of your watershed. This will give you a comprehensive view that you just can't get from other field inventory methods.

Some Extra Tools & Techniques of the Trade

In addition to documenting what you see on your watershed map and on Data Sheet 1, you may want to keep a narrative account or journal of your watershed field observations in a notebook. Photographs are another great tool for your watershed inventory; they can be very effective in documenting problem spots and changes that occur over time.

Where Do You Start?

You might want to begin your inventory at the headwaters of your stream and finish at its mouth. As you do so, recall the discussion on the **river continuum** in Chapter 1. Observe how different characteristics of the stream change, such as size, degree of meandering, and size of the materials that make up the stream bottom. Note these general observations on your watershed map. Make observations of the different items listed on Data Sheet 1 along the way. The procedures below correspond to those items:

1. See what the landscape in your watershed can tell you about **climatic patterns and events**. Look for evidence of flooding, such as damage to property, scoured out stream channels, eroded banks, and fresh sediment in flood plains. Also look for signs of drought, such as dry streambeds, wetlands, and low water levels. Talk to property owners about their experiences with flooding and drought.

2. Get a sense of the **topography** of your watershed. Is it steep or flat in the headwaters? Visit the highest elevation if possible, and also the mouth of your stream or river (the lowest elevation). Note how the topography changes from headwaters to mouth.

3. Visit some of the predominant **geological features** in your watershed , such as mountain peaks, ridges, hilltops, valleys, canyons, etc. Look for rock outcrops and other places where different geological layers may be exposed. Take samples of different rock types to identify, recording the locations where you found them. Keep an eye out for evidence of recent geological activity such as landslides.

4. Inventory the **water resources** in your watershed. If possible, visit the headwaters of your stream. Make trips to the tributaries that

feed into your stream, as well as to lakes and wetlands throughout the watershed. As you travel about, keep in mind where the aquifers are located. Visit any springs, test wells, or other places where you can see the groundwater.

5. Examine the different **soil types** that occur in your watershed. Note where hydric soils occur, as well as fertile soils that are suitable for agriculture and those more suitable for development. Make sure you document any evidence you see of soil erosion problems.

6. Inventory the **vegetation** in your watershed. Observe the different plant communities throughout your watershed from headwaters to mouth, as well as from streamside to upland. Note the plant species, native and introduced, that predominate in the different types of plant communities. Keep an eye out for endangered or threatened species if they are suspected to be in your watershed.

7. As you travel through your watershed, watch and listen for signs of **fish and wildlife**. Use the information you gathered from your local fish and wildlife biologists about key habitats and times of the year to look for different species. Visit any fish migration barriers, fish ladders, and fish hatcheries in your watershed.

8. Look for **historical features** in your watershed, such as old homesteads, farms, forts, and mining sites. Don't forget to visit significant Native American sites, such as fishing and plant gathering areas, or tribal village sites and burial grounds. Also, make a point to talk to watershed residents, especially those that have lived in the area a long time. You're likely to find a wealth of information in their accounts of what the watershed used to be like.

9. Observe **land use** in different parts of your watershed. Visit urban and rural areas. Include in your travels residential, agricultural, commercial, and industrial areas, as well as any areas that are managed for forestry, mining, recreation, and open space. Get a visual sense of the amount of impervious surfaces throughout your watershed.

10. Inventory **water use** in your watershed. Visit the sources that supply domestic, industrial, and irrigation water. Check out any test wells that federal, state, or local agencies may have constructed to conduct groundwater studies. Visit dams used for hydroelectric power generation or other purposes. If residents in your watershed are serviced by a sewer system, visit the treatment plant and locate the outflow. If residents use septic systems, visit those areas and keep an eye (or nose) out for possible trouble spots.

11. If your stream flows through more than one **political jurisdiction**, note the borders as you travel across them, and look for differences in land uses and in how aquatic resources seem to be treated. If you've spent some time talking to your local government agency staff and/or elected officials, you may have a good idea of issues to investigate.

12. As you travel through your watershed, take note of any potential problem areas which may affect water quality, the hydrology in your watershed, or fish and wildlife habitat. Don't forget that recreational activities may also have impacts, in addition to urban development, industry, farming, and forestry. Visit areas that are prone to flooding or stream segments prone to drying up.

NOW WHAT?

Now that you've got an overall sense of what's going on in your entire watershed, you're ready to head into Chapter 3, which will start you out on the process of looking more closely at the stream itself. The information you gathered in your watershed inventory will help you establish your stream monitoring goals and identify those portions of your stream that need particular attention. Update your watershed inventory once a year to keep track of how it changes over time.

Miss Mayfly's Guide To Watershed Etiquette, Safety and Liability

Design and Etiquette Guide by Kate O'Laughlin

Watershed Etiquette

• Don't drive off the road in your watershed. Driving off road in the woods or along a stream disturbs wildlife and degrades their habitats. If you want to see part of your watershed more closely, park your car in a lot or on the side of the road and explore by foot.

• While on foot, use the trails and paths in your watershed. They help keep you from getting lost and help the wilderness stay wild.

• Get a landowner's permission if you will be crossing his or her property.

• Walk through your watershed quietly. Listen to the sounds of wildlife residents; disturb them as little as possible.

- Leave your dog at home. Dogs can disturb or injure wildlife and in turn, can be injured by other animals.

- Take care not to disturb or alter wildlife habitat. A dead tree trunk or hole in the ground may be home to many kinds of wildlife.

- Try to minimize the impacts on the stream and its fragile stream banks. Breakdown of stream banks increases erosion and adds excessive sediment to a stream. If you are surveying or monitoring a stream with a group, only those people actually making the measurements should be in the stream or on its banks. Choose access points to the stream that cause the least disturbance.

- Miss Mayfly says: "Take only pictures; leave only footprints." Make observations and document evidence of wildlife, plants, and other organisms, but don't remove eggs, nests, organisms, or whole plants.

- If you feel the need to sacrifice plant or macroinvertebrate specimens for better identification, please do so sparingly. Plants provide food and shelter for wildlife and macroinvertebrates. Macroinvertebrates are food for fish, birds, frogs, and salamanders.

- Put any rocks and woody debris you remove back into the stream. They are habitats for fish, bugs and plants.

- Know what kinds of fish live in a stream and their life histories. Know when they spawn and when young fish emerge from the gravel streambed. Fish are most vulnerable at these times and they should be disturbed as little as possible.

- Take your trash back with you for recycling. If nature calls, locate your bathroom "facilities" at least 100 yards away from a stream or lake. Miss Mayfly says: "Dig a hole, bury your wastes, and bring your toilet paper back with you." (A plastic bag works well).

(continued on next page)

You may want to make a plant press if you find it necessary to preserve some of the plants from your watershed for identification and presentation purposes. Remember to sample sparingly to lessen your impact on the natural environment.

Watershed Safety

- Try not to work alone. Work with a "buddy" who can pull you out of the mud or go get help if you fall down twist an ankle. An added bonus is that two sets of eyes and ears are better than one.

- Always let someone back at home, office, or school know where you're going and when you're due to return. Plan to finish your work well before dark.

- Be prepared for unpredictable weather before you head out into your watershed. Always take along rain gear and appropriate clothing. A sudden storm may blow in, you might fall into your stream, or you may be out longer than you expected. A change of warm clothing and footwear is a wise precaution. Be wary of hypothermia.

- Watch for dangerous objects such as sharp rocks, broken glass, steep slopes, slippery logs, thorny plants, and stinging insects. Sturdy footwear and long pants are recommended.

- Take care when wading in streams. Try to stay at knee depth or less if possible, and stay out of dangerously swift currents. Use a staff or other cane-like device to help support your balance in the middle of the stream. Don't let your waders fill up with water!

- Don't drink the water from a stream unless it is filtered and treated. Though the water quality may appear to be excellent, microscopic organisms normally living in streams can cause severe flu-like symptoms in humans.

- If you suspect a stream is highly polluted, wear protective rubber gloves if you need to put your hands in the water. Always wash your hands thoroughly when you return from the field.

- It's a good idea to become familiar with orienteering techniques. Take your maps and compass and learn how to use them. There are numerous books on the topic, but you can get a great "starter" publication, entitled *Orienteering* from the Boy Scouts of America for about $1.50.

Trespass and Liability Issues

- If your watershed inventory requires entry onto privately owned land, make sure you know who owns the land and gain their permission before you proceed. This may be accomplished via a written "right of entry agreement" which clarifies for both parties what actions will be taken, who will be doing the work, when it will be conducted, and for what duration. If you enter private property without permission, you are trespassing and are liable for any problems that arise.

- Whenever possible, form a "partnership" with the land owner. Inform them of the purpose of your study, who you'll be sharing the information with, and the potential outcome of your efforts. If you feel their management practices may be harming the stream, try to understand their point of view before you start to make suggestions.

They may have some good reasons for their activities; listening will help you come up with realistic solutions that can work for both of you. Try to aim for cooperative "win-win" solutions. See chapter 10 "Putting Your Data to Use," for more information on this topic.

- If you and your fellow Streamkeepers are collecting information for local, state, provincial, or federal agencies, ask them if they will provide insurance for your and your group menbers. Large service organizations may be willing to insure you as well. Short of being insured, the next best step is to prepare an "acknowledgment of risk" form describing the potential hazards of your project; then provide your fellow Streamkeepers with a thorough safety briefing ending with everyone's signature on the form. In short, cover your behind!

Date _____ Name_____Group_____

Watershed Name_____ USGS quad(s)_____

Begins in_____ Flows Through_____

Ends in_____ (name towns, counties, states, regions, etc.)

Drains into_____ (name body of water)

Square Miles_____ Approx. Length_____ Width_____

Driving/Hiking Directions_____

_____

CLIMATE

Average yearly precipitation   < 10 in ❏   10-50 in ❏   50-100 in ❏   > 100 in ❏

Most of the precipitation is in the form of   rain ❏   snow ❏   other ❏ _____

Most precipitation occurs in (month(s))_____Floods most commonly occur in (month(s)) _____

Droughts most commonly occur in (month(s), year(s))_____

Coldest month of year_____ Warmest month of year_____Yearly temp. range _____

GEOLOGY / TOPOGRAPHY

Describe briefly the geologic history that shaped your watershed _____
_____
(add separate sheets as necessary)
Describe the physical characteristics of different reaches of your watershed:

	Upper reaches	Middle Reaches	Lower Reaches
Uplands (mountains, hills, or flat)	_____	_____	_____
Valley (broad, medium, narrow)	_____	_____	_____
Gradient (steep, medium, gentle)	_____	_____	_____
Channel (straight, meandering)	_____	_____	_____
Bottom (boulder, cobble, gravel, fines)	_____	_____	_____

Predominate rock types   igneous ❏   sedimentary ❏   metamorphic ❏

Specific rock types present_____

Highest point_____ Lowest point: _____
(include elevation and location)

Geologic Activity: earthquakes ❏   volcanic eruptions ❏   landslides ❏   other ❏ _____

WATER RESOURCES

Headwaters originate from   glaciers ☐   snowmelt ☐   rain ☐   wetlands ☐   lakes ☐

Length of your stream _____   groundwater ☐   springs ☐

Names of tributaries_____

Order of your stream at the outlet point of your watershed _____

Names of lakes _____

Number of wetlands   few (1-15) ☐   many (15-25) ☐   Abundant (>25) ☐

Areas underlain by aquifers (if any) _____

SOILS

Predominate soil types_____

Areas with soil suitable for farming_____

Areas with soil unsuitable for development_____

Areas with potential soil erosion problems_____

VEGETATION

List the native and introduced plant species that dominate the different plant communities of your watershed:

	Native	Introduced
Upland forest	_____	_____
Riparian	_____	_____
Grassland	_____	_____
Other plant communities	_____	_____

Percent of your watershed now covered by native vegetation _____%

Reasons for the loss of native vegetation _____

Time period over which the loss occurred_____

Endangered or threatened plant species _____

FISH

Native species (circle if endangered or threatened)_____

Non-native species_____

Locations of fish hatcheries and species produced_____

Types and locations of barriers to fish migration_____

WILDLIFE

Native species (circle if endangered or threatened) _____

_____

Non-native species _____

Key wildlife habitat areas _____

HISTORICAL (Attach extra pages as needed)

The earliest human inhabitants were _____

Describe briefly the settlement of your watershed _____

_____

Cultural and historical resources in your watershed _____

_____

DEMOGRAPHICS

Current watershed population _____ Projected population in 10 years _____   20 years_____

Watershed population 10 years ago _____   50 years ago _____   100 years ago _____

Areas where most of the people live _____

List towns, cities, & counties and the percent of watershed land area that each encompasses

_____
            (attach list of elected officials for each jurisdiction)

What makes people want to live (or not) in your watershed ? _____

LAND & WATER USES

Estimate the percent of your watershed zoned for each land use and check the activities that apply

rural residential _____%   densities (# of houses per acre)_____

urban/suburban residential _____%   densities _____

commercial _____% light commercial ❒   heavy commercial ❒

industrial _____%   light industry ❒   heavy industry ❒

agricultural _____%   grazing ❒ confined animal ❒ dry crops ❒   irrigated crops ❒ nursery ❒

forestry _____%   clear-cut ❒   selective cut ❒   roads ❒   tree farm ❒

mining _____%   type of mining: _____

DATA SHEET 1 WATERSHED INVENTORY

parks/open space _____% swimming ❑   boating ❑   fishing ❑   other ❑ _____

other recreation _____%   ski resort ❑   golf course ❑   other ❑ _____

Percent of watershed that is   public land _____%   private land _____%

Percent of watershed covered with impervious surfaces _____%

Sources of domestic water supply for watershed residents _____

Location of sewage treatment plants (if any) servicing watershed residents _____

Areas that rely on septic tanks _____

Altered hydrology (dams, diversions, detention systems, culverts, dikes, drained wetlands, etc.)

type of alteration	location	purpose
_____	_____	_____
_____	_____	_____
_____	_____	_____
_____	_____	_____

(add lists as necessary)

WATER QUALITY / QUANTITY CONCERNS

Water quality classification(s) of your stream and tributaries _____

   (attach standards and regulations protecting watershed resources, include enforcing agencies)

List pollutants of concern and their potential sources (include locations if possible)

pollutant	point source	nonpoint source
_____	_____	_____
_____	_____	_____
_____	_____	_____
_____	_____	_____

AREAS PRONE TO FLOODING/DRYING UP (Add extra pages if necessary)

Location	Circle One	Dates
_____	Dry Flood	_____
_____	Dry Flood	_____
_____	Dry Flood	_____

The Streamkeeper's Field Guide

PART TWO

Your Stream's Health

## CHAPTER THREE

Monitoring Program Design

Now that you've involved yourself in the business of watersheds, you're ready to focus more specifically on the stream. You know that a stream can be viewed as a narrow but very active strip of water which drains the watershed, a much larger surrounding land area. And you know that because runoff to streams originates from precipitation all over the watershed, whatever happens on the land is reflected in streams. When we watch them carefully, streams serve as barometers of watershed health.

When designing a monitoring program, it is helpful to consider the following questions:

- **Why** do you want to monitor the stream?
- **What** parameters will you monitor?
- **How** will you monitor those parameters?
- **Where** will you collect your information?
- **When** will you collect your information?

THE "WHY" QUESTION

It is of utmost importance to have a clear idea of why you are interested in monitoring the stream, or in other words, what your overall goals are. Once you answer this question, all else will follow; knowing why will help you figure out what, how, where, and when.

What you discover while conducting your watershed inventory should help you formulate your monitioring goals. You may have found a concern in the community about whether a stream is safe for domestic use. Perhaps the stream is a drinking water supply or is used extensively for recreation. In this case, you would monitor water quality parameters to see if the stream meets standards set by the state for these different human uses. Monitoring a water body to determine if it meets established water quality standards is called **compliance monitoring**.

For many Streamkeepers, an interest in streams comes from a concern for fish and wildlife. Perhaps in your watershed inventory you find that a particular fish population is experiencing a serious decline. Since most states set water quality standards for fish and other aquatic life, you can conduct compliance monitoring to determine if the stream meets those standards. However, you would also want to make sure your monitoring program includes parameters relating to the habitat and food sources required by that fish species, in addition to water quality parameters.

If you found there is a paucity of information about your stream, you may decide to conduct a **baseline study** to characterize the overall health of the stream. This will enable you to document existing conditions and establish a data base for planning or future comparison. If your stream is in good shape but is threatened by a proposed land use, you can use your baseline data to compare with changes that might occur in the future. It also gives you the information needed to determine if there are any potential problems that need further study. This is **trend analysis.**

Perhaps your watershed inventory led you to suspect that a particular land use is adversely affecting your stream. In this case, you might want to conduct an **impact analysis** of that potential problem, to determine whether it is indeed a problem, and to what degree. The type of land use activity will dictate what parameters you analyze. For example, if you are assessing the impact of forest practices, you would include in your study an analysis of the stream bottom, because forest practices tend to result in increased sedimentation.

Perhaps you find evidence that a certain land use is responsible for a problem in your stream, and you are able to develop a program aimed at correcting the problem. Maybe you have worked with a land owner to implement **best management practices (BMP'S)** to prevent further stream degradation. In this case, you could design a study to assess whether the BMP's have had a restorative effect.

Your goals will evolve as you delve into the process of becoming a Streamkeeper. As you become familiar with your watershed and get to know your stream, you will discover new things that will help you continue to refine your goals, revise your plans, and find new parameters to monitor.

Whatever you decide your overall goals are, it is important that you collaborate with any private groups and government agencies that are interested in or have jurisdiction over the stream in question. Find out what their goals are and determine how your goals relate to theirs. If a certain group or agency is already collecting data on the stream, you can find out how your efforts can best complement theirs without "reinventing the wheel."

You also want to involve the resource users in the watershed, i.e., businesses, land owners, residents, and even visitors, because everyone in the watershed affects the stream. Getting potential polluters involved in your goal-setting process can make allies of those who you thought might be enemies. The most successful streamkeeping efforts are collaborations representing all watershed players.

A Plug for a Holistic Approach

The Adopt-A-Stream Foundation encourages Streamkeepers to follow a **holistic approach** to stream monitoring, or in other words, to be concerned with the overall health of a stream. Adopting a holistic approach means not only considering the chemical aspects of the stream's water quality, but the physical and biological character of the stream as well. It means going beyond focusing on a particular threatened species, to looking at the different communities of plants and animals within the stream ecosystem.

Until recently in the U.S., the Federal Clean Water Act focused exclusively on the chemical composition of the Nation's waterways. Given this criteria, water in a pipe could rank as high (or higher) than a stream in a water quality evaluation. We have found that stream systems can be technically "clean" from a water chemistry perspective and yet be a biological desert! Now, the U.S. Environmental Protection Agency is being asked to consider the physical, chemical, and **biological integrity** - the living resources - of the nation's waters.

Dr. James Karr of the University of Washington has developed a model that considers five factors contributing to the biological integrity of surface water systems. These factors include water quality, habitat structure, energy (nutrient) sources, flow regime, and biotic interactions. Karr advises Streamkeepers to look directly at the biological community of their streams. If it is degraded or "out of sync," you should be able to attribute the imbalance to the degradation of one or more of the five factors. A diagram of Dr. Karr's five factor approach is on page 46.

With this holistic approach, you will have a more comprehensive view of your stream, and thus find solutions that treat the causes, rather than the symptoms, of your stream's problems. For example, artificially introducing fish into a stream system to bolster the population will not be effective if the population decline is caused by a lack of sufficient food supply. Perhaps the aquatic insect populations are "on the brink" because of a loss of streamside vegetation and increased sedimentation from unstable stream banks. In this case, restoring the vegetation and controlling the sediment problem are better solutions than restocking the stream with more hungry fish.

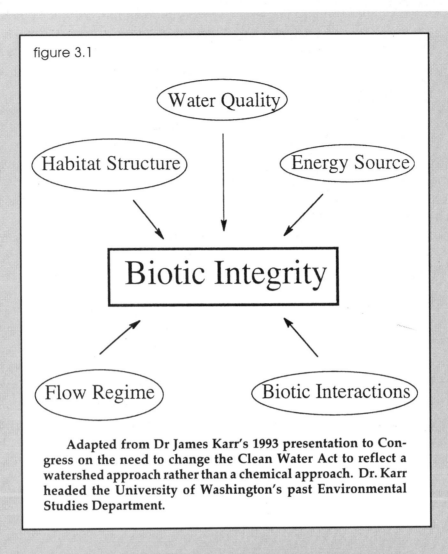

figure 3.1

Water Quality

Habitat Structure

Energy Source

Biotic Integrity

Flow Regime

Biotic Interactions

Adapted from Dr James Karr's 1993 presentation to Congress on the need to change the Clean Water Act to reflect a watershed approach rather than a chemical approach. Dr. Karr headed the University of Washington's past Environmental Studies Department.

Water Quality — the extent to which a stream strays from its natural chemical makeup due to human impacts.

Energy Sources — the source of **Biotic** nutrients for the stream's food chain. These sources may be: natural, such as decaying leaves and woody debris, or human caused, such as sewage and fertilizer runoff.

Biotic Interactions — the links between species in the **food chain** (or simply who's eating who), and an awareness of the impacts on one species when another species disappears. For example, caddisflies, mayflies, and stoneflies are primary food sources for salmon, steelhead, and trout. If these insects are not present in your stream, those fish dependent on them will not be present.

Flow Regime — the volume of water flowing through a stream over time. Changes in flow regimes and attendant impacts on fish and wildlife must be evaluated on par with flooding, erosion, and municipal/industrial/agricultural uses of water.

Habitat Structures — the types and amounts of natural features that provide fish and wildlife habitat structures, e.g. in-stream logs, pools and riffles, and undercut banks for fish; and abundant and diverse vegetation for wildlife.

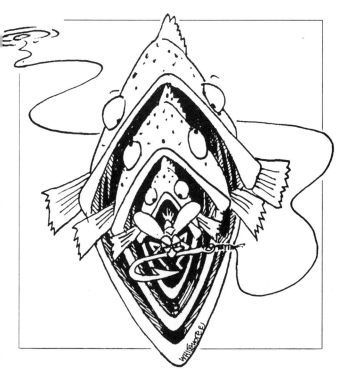

THE "WHAT" QUESTION

Once you have formulated the goals of your stream monitoring program, you will be able to decide which parameters to monitor. The parameters included in *The Streamkeeper's Field Guide* are based on the goal of gathering baseline data on the overall health, or biotic integrity, of a stream. The guide was written in this way to encourage Streamkeepers to adopt a holistic approach to stream monitoring. In additon, the comprehensive array of parameters required by such an approach also provide Streamkeepers with a monitoring menu from which they can pick and choose to suit their own goals.

Detailed background information and procedures on the parameters listed below can be found in Chapters 4-7 of this book:

- **fish and wildlife:** finny, furry and feathery creatures inhabiting the stream and surrounding area
- **benthic macroinvertebrates:** spineless critters that inhabit the stream bottom which are large enough to see with the naked eye
- **vegetation:** plants in the area along the stream
- **overhead canopy:** extent to which trees shade the stream
- **stream channel gradient:** steepness of the stream

- **stream channel sinuosity:** longitudinal or meandering pattern of the stream
- **stream valley and channel shape:** cross-section shape of the valley the stream flows through, and the cross section shape of the stream channel itself
- **stream banks:** extent to which banks are stable, eroding, or altered
- **instream habitat:** presence of pools, riffles, runs, glides, boulders, large woody debris, protective cover, etc. (see Chapter 4 for a definition of these terms).
- **substrate:** nature and condition of the stream bottom
- **flow and precipitation:** volume and velocity of water flowing down the stream, as well as the amount of recent precipitation in the watershed
- **chemical water quality parameters:** many parameters, including pH (how acidic or alkaline the water is) and the concentration of dissolved oxygen(DO) in the water.
- **temperature:** temperature of the stream water, and the temperature of the air
- **land uses:** activities in the area surrounding the stream
- **human alterations to the stream channel:** human activities affecting the character of the stream channel, instream habitat, substrate, flow of water, water quality, etc.

Each parameter has a different sensitivity to human activities and will tell a part of the story. For instance, the stability of stream banks will probably not be affected directly by use of fertilizers by a farmer, but they will definitely be affected by grazing. To examine the effects of both of these land uses, it is necessary to look at water quality parameters as well as the condition of the stream bank.

For those of you who want to focus on assessing the impact of a particular land use on a stream, Table 3 will help you choose parameters. This table links the effects of specific land use practices to problems that may appear in streams as a result. The table also recommends specific parameters that are most effective to test in order to make an appropriate diagnosis.

THE "HOW" QUESTION

While they are comprehensive, the procedures presented in this book are fairly simple. The guide was written with the assumption that most Streamkeepers do not have access to expensive equipment or laboratories. The procedures are designed to be carried out in the field with relatively inexpensive, easily obtainable, or homemade equipment.

For example, the flow measurement procedure requires an orange, a stopwatch, a measuring tape, and a homemade measuring stick, not a $1000 flow meter. Collecting benthic macroinvertebrates requires use of a homemade kick net and ice cube trays, among other easily scrounged implements. Measuring pH and dissolved oxygen can be done with $40.00 field test kits as opposed to expensive meters.

Although these methods may be less precise than those used by agency scientists, they can still provide valuable information. In fact, if you follow all the procedures outlined in this book, it's likely that you'll have a more comprehensive study of a stream than many agency scientists have produced.

Once you have delved into the world of stream monitoring, you may decide you are ready for a higher level of analysis. You may decide to monitor fecal bacteria, which requires the use of a laboratory incubator. Or you may decide you need a lab with stereoscopes in order to analyze your stream's macroinvertebrate population in more detail.

Try not to let funding issues keep you from reaching your goals. There are ways to gain access to the high tech equipment and laboratories that don't require a capital investment, such as creating a partnership with a school, agency, university, or other entity that already has the resources you need.

The Streamkeeper "Quality Assurance Plan"

No matter what parameters you measure and which procedures you follow, the best way to ensure the quality of your data is to develop a **quality assurance plan**. A quality assurance plan documents the answers to all the questions you considered in designing your stream monitoring program. It is a blueprint that lists your goals and what parameters you are monitoring. It outlines your methods, the type of equipment you are using, where and when you are collecting information, as well as who is conducting the monitoring. A quality assurance plan also documents how you **calibrate** your equipment, a procedure used to ensure accuracy.

Streamkeepers are encouraged to get a three ring binder notebook and construct a thorough quality assurance plan. Follow the steps outlined in Chapter 8, "Credible Data." Having a quality assurance plan will make it easy to train new people joining your Streamkeeper group. Using a systematic approach will reduce the chances of errors and increase the validity of your information. With credible information in your back pocket, you'll be able to influence decision makers to correct problems you discover in your stream.

TABLE 3 LAND USES AND POTENTIAL SYMPTOMS IN STREAMS

LAND USE & ACTIVITIES	POSSIBLE EFFECTS ON THE WATERSHED	POINT & NONPOINT POLLUTANTS	DIAGNOSIS & SYMPTOMS
FOREST PRACTICES road building timber harvest tree farming	decreased slope stability removal of trees and streamside vegetation removal of understory vegetation loss of water storage capacity increased erosion increased runoff	sediment fertilizers (nutrients) herbicides	eroded stream banks greater proportion of fine sediments on stream bottom murky water (increased turbidity and suspended solids) increased temperature decreased dissolved oxygen macroinvertebrate population dominated by types that feed on fine sediments and do not require high levels of oxygen increased streamflow fluctuations increased nutrients (nitrates and phosphates) increased plant and algal growth increased herbicides (e.g. 2,4-D) and pesticides (e.g. toxaphen, aldrin, chlordane)
AGRICULTURE livestock crops	soil compaction and loss of water storage capacity manure removal of streamside/ wetland vegetation increased erosion increased runoff	sediment nutrients herbicides pesticides	same as above, plus increased fecal bacteria
URBAN/ INDUSTRIAL Urban water and waste transportaion (sewage, septic tanks, stormwater runoff) Industrial processes, (including mines and pulp mills) Home/community (including construction)	loss of vegetation increase in paved areas increased runoff increased erosion increased sediment and debris	sediment nutrients human wastes fertilizers (nutrients) herbicides pesticides heavy metals toxic substances	same as above, plus stream may be scoured down to hard bedrock murky water (increased turbidity and suspended solids) precipitate on stream bottom water appears oily, foamy, or with odd color water smells like rotten eggs, fuel, soap, etc. increased fecal bacteria presence of heavy metals presence of petroleum compounds presence of detergents

THE "WHERE" QUESTION

Unless you are working on a very small stream or have unlimited time, people power, and resources, it will be impossible to monitor the entire length of your stream. You may be able to cover the entire stream, and its tributaries, for the more general observations that are part of your watershed field inventory. But for your more detailed monitoring activities, you will have to select sections of your stream to evaluate. Each section that you evaluate is called a **stream reach**.

If you ask a biologist what a stream reach is, you may hear, "it's where there is a change in habitat structure in the stream." A geologist may say, "A reach begins and ends where there are obvious breaks in topographical gradient." Both perspectives are useful; however, the biologist and geologist may have different opinions on how to divide a stream

into different reaches. For the sake of consistency and simplicity, we recommend that you define a stream reach as a 500 foot long section of your stream. 500 feet is long enough to provide sufficient data, and having a consistent length will make it easy to measure and find your reaches.

The actual locations you choose for your 500-foot long stream reaches will depend on the goals of your study. This is where you will want to consider the biologist's and geologist's perspectives, and recall what you learned in Chapter 1 about the river continuum and different stream orders. Below are recommendations that will help you select reaches if you're planning to conduct a baseline study of your stream's condition, or if you are planning to assess the potential impact of a particular land use on your stream.

Baseline Study

If you are conducting a baseline study of the stream, you may want to select reaches that represent different stream orders, to obtain a complete picture of the entire stream system. You might select one reach in a headwater tributary, another reach in a second or third order section of the stream, and a third reach close to the mouth of the stream.

If you get overwhelmed in an effort to obtain a complete picture of your stream, perhaps you are focusing on a stream system that is too large. If this is the case, consider reducing the scope of your study to a smaller tributary or segment of your stream. Or, better yet, recruit other Streamkeepers to help you cover more ground.

figure 3.2a
Watershed

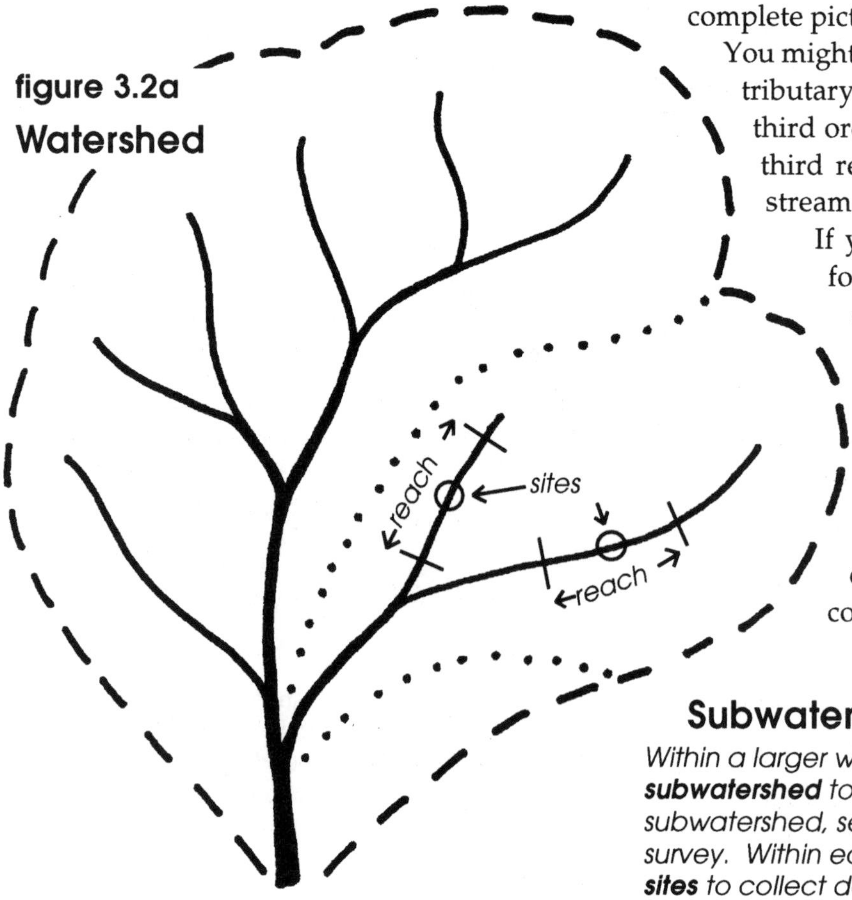

Subwatershed

*Within a larger watershed, select a **subwatershed** to inventory. Within the subwatershed, select **stream reaches** to survey. Within each stream reach, select **sites** to collect data on the stream bottom, flow, macroinvertebrates, and water quality.*

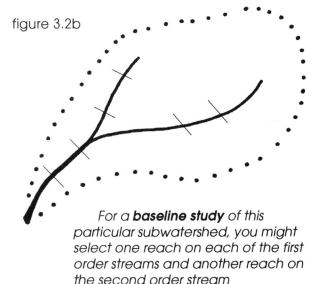

figure 3.2b

*For a **baseline study** of this particular subwatershed, you might select one reach on each of the first order streams and another reach on the second order stream*

Impact Assessment

If you are assessing some type of potential human impact on a stream, you should compare two reaches, one immediately upstream, and one just downstream, from the alleged impact. In addition, select a third reach that is further downstream from the impact to evaluate the degree to which the stream has recovered from the upstream condition.

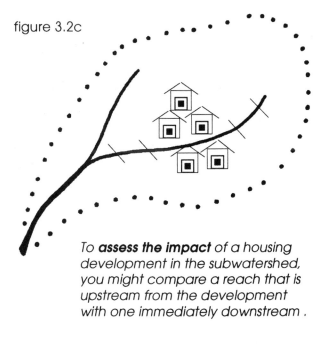

figure 3.2c

*To **assess the impact** of a housing development in the subwatershed, you might compare a reach that is upstream from the development with one immediately downstream.*

For an impact assessment, it is extremely important to select stream reaches that minimize complicating factors. Be careful not to blame a particular land use for a conditon in your stream that is caused by natural phenomena or some other event.

For example, the lower gradient reaches of a stream are often naturally silty-bottomed, providing habitat for some of the macroinvertebrates that tend to be quite tolerant of organic polution. Do not assume that a low gradient section of your stream is degraded by a land use just because it is silty and full of aquatic worms. This may be its natural condition.

How does one sort out these complications? Make sure your upstream and downstream reaches are comparable, that you're not "comparing apples with oranges." The upstream reach is your **reference** reach, which means that in the absence of the alleged impact, all other characteristics of your downstream reach should be similar to the upstream reach.

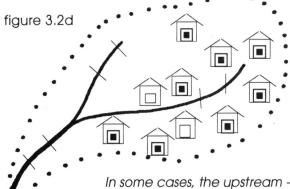

figure 3.2d

*In some cases, the upstream - downstream comparison may not be possible. For example, if a stream is impacted along its entire length, there would be no appropriate upstream reach to act as a control. In this case, you might compare the impacted reach to a **reference** reach on a similar, but unimpacted, stream.*

The two reaches should be in the same place in the river continuum, with similar gradient, substrate, valley shape, channel size, etc. A reach with a substrate of bedrock should not be compared to one with a substrate of cobbles, nor a low gradient, meandering stream to a high gradient, white-water torrent, nor a wadeable stream to one that requires a boat for sampling.

Also, there should not be anything between the two reaches that significantly affects the hydrology, such as a new tributary or a beaver dam. If you're trying to assess the impact of a single land use, no other land uses should be present.

Unfortunately, such ideal conditions for assessing an impact are hard to find. If you run up against this problem, another option is to establish reference conditons from historical data on the stream. If you do this, make sure your monitoring methods are comparable to the methods that were used to collect the historical information.

If adequate historical data is not present, try to establish reference conditions from other streams, such as nearby tributaries in the same watershed, or streams in a neighboring watershed. A stream used to establish reference conditions should be in its natural, undisturbed state, and not downstream from sources of contamination. Make sure it has similar gradient, substrate, valley shape, and channel characteristics as your stream. It should be of the same order as your stream, with a similar volume of flow.

Stream Reaches and Study Sites

Once you've chosen the general location for your stream reaches, you still have to decide at what exact point they will begin. The beginning of a reach should be at a location that can be easily found by anyone for repeat visits. It can be a spot with a significant landmark, such as a road crossing over the stream, a property line, or a large tree.

Some of the parameters you'll be monitoring are evaluated along the length of your stream reaches. Chapter 4 will guide you through a stream reach survey, in which you collect information about parameters such as streamside vegetation, gradient, instream habitat, and land uses.

Other aspects of your stream, such as the stream bottom, flow, cross section, macroinvertebrates and water quality parameters, are evaluated at specific study sites within your stream reaches. Thus, once you've selected your stream reaches, you will need to select study sites where you will evaluate these parameters. Information on how to select study sites is presented in chap-

ters 5-7, along with the instructions for monitoring those particular parameters.

By conducting investigations of stream reaches and study sites, it becomes possible to learn a great deal about the stream as a whole. A reach and its study sites will be a barometer of stream health, like a stream is to a watershed. If the stream is a picture of good health, the reach and its sites will be in good shape; if there are problems in the stream, the reach and sites will reflect those problems.

THE "WHEN" QUESTION

In general, for those parameters surveyed along the length of your stream reaches, we recommend evaluating them once each in the fall, winter, spring, and summer, to get a sense of seasonal variation that occurs. We also recommend a quarterly survey of the stream bottom and a quarterly or biannual survey of macroinvertebrates. As with the other questions, the answer depends on your particular goals and resources.

If you only have time and resources to survey your stream reach, stream bottom, and macroinvertebrates twice per year, you'll still end up with plenty of information about your stream. The most important times to monitor macroinvertebrates are: 1) late spring after snowmelt but before the leaves are fully out and 2) early autumn before leaf fall. It is important that you conduct your stream reach and stream bottom surveys at the same time you collect macroivertebrates.

For the parameters that fluctuate more often, such as flow, temperature, and chemical water quality parameters, more frequent monitoring is recommended. Depending on your concerns, you may want to collect information on these parameters monthly, weekly, or even daily.

For example, if your stream suffers from high temperatures and low oxygen during the summer, you may not want more than a week to slip by between measurements. You may even want to take measurements twice a day, for a limited time, to pick up daily fluctuations.

See Table 4 for a summary of when and how often to monitor different parameters. For certain

TABLE 4   A SUMMARY OF "WHEN, "WHAT," AND "WHERE"

When	What	Where
once per year	watershed inventory	entire watershed
quarterly	stream reach survey	within 500' x 500' areas along your stream
quarterly	cross section profile plot	study sites within your stream reaches representing different habitat types and/or problem areas
quarterly	substrate	study sites within your stream reaches representing different habitat types and/or problem areas; at same sites chosen for macroinvertebrate collection
biannually	macroinvertebrates	study sites within your stream reaches representing different habitat types and/or problem areas
monthly	flow & precipitation	at an appropriate site within your stream reach
"	temperature	near macroinvertebrate site
"	chemical water quality parameters	near macroinvertebrate site
additional monitoring; weekly, daily, or hourly, as necessary	flow & precipitation temperature chemical water quality parameters	near macroinvertebrate site; also at potential problem areas, if any

combinations of parameters, it is important that you monitor them at the same time.

For example, every time you sample your stream's macroinvertebrate population, you also should survey the stream bottom and measure the flow, temperature, and chemical water quality parameters. Then you can use all the information to help sort out the factors affecting the macroinvertebrate population. More information is provided in Chapters 4-7 with the detailed instructions on each parameter.

How Long Should Your Study Last?

We recommend you conduct baseline monitoring for a minimum of two years because of the natural variation that occurs over time in stream systems. While pollutant-bearing runoff is high in times of heavy rainfall, pollution problems may be even greater during periods of low rainfall because reduced flow increases their concentrations. The longer you monitor, the better you will understand the effect of seasonal changes on your stream.

Because natural seasonal variation might exacerbate or conceal the effect of human activities, we also recommend you conduct your impact assessment studies over a long enough period of time for you to gather conclusive evidence of a suspected problem. This may range up to two years, depending on the nature of the problem. As with baseline monitoring, the longer you collect information, the better you will understand what is going on in your stream.

If your Streamkeeper group does not have the resources or time available for long term monitoring, you can still play an important role in protecting and restoring streams. Short term monitoring can serve a "watchdog" function for other groups or government agencies by keeping an eye out for problems that need further investigation. Moreover, it may not take too much investigation to find some of the more obvious problems that can be fixed. There are many degrees of Streamkeeper involvement; find the commitment level that is right for you and your group.

CHAPTER FOUR

The Stream Reach Survey

At some time in your life, you may have marveled at a salmon leaping up a waterfall on its way to a spawning bed. You may have been thrilled by catching your first "striper" bass, or catfish. Or you may have simply felt content sitting on a rock as a school of minnows passed by, knowing your stream was alive. Indeed, the joy you experienced at your local fishing hole may be why you became "hooked" on becoming a Streamkeeper in the first place.

And now you know how to design a stream monitoring program. At this point you've set your goals, decided on the parameters you will monitor, selected the stream reaches from which you will collect your data, and have a sense of when, how often, and for how long you will be

Illustration by Sandra Noel from *Adopting a Stream: A Northwest Handbook,* published by The Adopt-A-Stream Foundation, 1988.

evaluating the stream. You're ready for the first actual monitoring step - assessing the condition of the stream reaches that you have selected. By conducting the **stream reach survey** you will learn a great deal about the fish and other creatures that inhabit your stream.

The stream reach survey involves walking the length of the 500' reach along the stream channel, evaluating various parameters as you make your way from the beginning to the end of the reach. But as you know, a stream reflects what's going on in the surrounding landscape as well, so it's important to extend your surveying efforts "beyond the babbling brook."

This chapter begins with background information about the parameters included in the stream reach survey. Collectively, they are evaluated both along the stream channel and in the areas on either side of the stream. For some parameters, we recommend you survey an area that extends 250' from either side of the stream, in total a **500 x 500 square foot area**.

An effective way to document what you find is to create a detailed map of your stream reach. **Stream reach maps** can vary from simple sketches to accurate, professionally surveyed representations. By drawing maps, you can identify habitats for later reference, such as where to look for fish and wildlife, identify problem areas, and document changes over time. A field procedure that outlines detailed instructions for making a base map of your stream reach follows the background information section of this chapter.

The last part of the chapter contains detailed field procedures for the parameters included in the stream reach survey. In addition to documenting information on your stream reach map, you'll want to record information on Data Sheet 2, "Stream Reach Survey."

One of the first items to record on Data Sheet 2 is the location of your stream reach. It is also important to document the location of your stream reach in your Quality Assurance Plan. You cannot survey your reach over time with precision if you do not know exactly where it begins and ends. This chapter provides instructions on how to do this in the section entitled "Documenting Locations," on page 74.

Finally, before you venture into the field to map and survey your stream reach, make sure you consult "Miss Mayfly's Guide to Watershed Etiquette, Safety, and Liability" (pages 34-37) for reminders on how to least disturb your watershed and keep safe while monitoring.

Illustration by Sandra Noel; from *Adopting A Stream: A Northwest Handbook*, an AASF publication.

BACKGROUND INFORMATION FOR THE STREAM REACH SURVEYOR

The parameters covered by this section include: fish, wildlife, vegetation, overhead canopy, stream channel gradient, channel and valley cross section shape, channel sinuosity, stream banks, instream habitat, human alterations to the stream channel, and land uses adjacent to the stream reach.

Fish

Some simple field procedures for surveying fish in your stream are provided later in this chapter and in the appendices. In order to put these procedures to the most effective use, you will first need to understand the habits of your stream's fish, and even begin to "think" like a fish.

There are numerous families of fish that may inhabit your stream ranging from sturgeon (Acipenserdae), freshwater eels (Anguillidae), freshwater catfish (Ictaluridae), salmon (Salmonidae), lampreys (Petromyzonidae), sculpin (Cottidae), sticklebacks (Gasterosteidae), suckers (Catostomidae), sunfish (Centrarchidae), perch and darters (Percidae), herring (Clupeidae), minnows (Cyprinidae), and many more.

More than 75% of freshwater fish populations in the United States come from the following six families: minnows, suckers, catfishes, sunfishes, perch, and darters. These fish are restricted physiologically to fresh water. Other fish, like salmon, shad, and smelt are **anadromous**, living most of their life in salt water, returning to fresh water only to spawn. The American eel, on the other hand, is **catadromous**, spending most of its life in fresh water, then going to sea to spawn.

Within each family, there are several different species. For example, the salmon family (Salmonidae) includes salmon, steelhead, trout, char and Dolly Varden. Another favorite of fishermen are bass, which are part of the sunfish family. And there are hundreds of species of minnows which generally are quite small; however, carps, which are members of the minnow family, can grow to a length of three to five feet.

Fortunately, there are a lot of guide books and keys available in book stores to help you learn about the different types or "families" of fish that may inhabit your stream. A good general guide to add to your library is *How to Know the Freshwater Fishes* by Samuel Eddy and James Underhill. You can also obtain numerous scientific studies and detailed keys from local fish and wildlife agencies.

Guides, studies, and keys can be complemented by information from "old-timers" at your local fishing tackle store, who are generally out in the streams more than anyone else. Hard-core anglers, for example, can tell you the difference between a young salmon and a trout. (The trout's tail is usually moderately forked and even looks blunt. The salmon has a deeply forked tail.)

When drawing on these numerous resources, pay particular attention to the life cycle of the fish that you are observing. For example, adult salmon generally return to streams in the fall to lay eggs. The adults die. Juvenile salmon will hatch and begin to emerge into "fry" several weeks later and in the spring look like small minnows to the casual observer. Then, depending on the species, they will feed and grow in their stream from a few weeks to a year or more before migrating downstream to the ocean.

Of course, the life cycle and habitat needs of bass, catfish, stripers and other fish that may inhabit your stream are different. By understanding the normal migration patterns of the fish in your stream, you will have a pretty good idea which finny critters you should find when you begin your survey.

Also, by learning the habitat preferences of your stream's fish, you will have an easier time locating them. Not every fish likes cold, clear water with mixtures of pools and riffles that salmon and trout prefer. Remember that sculpin, dace, and carp have their place too. And the conditions and habitats that they prefer will tell you a story about your stream.

Wildlife

Brush piles and rock dens, feeding grounds and watering holes, dead trees with cavities, trails too small for people, scat, tracks, feathers, fur, and other "critter signs" may be present in and along your stream reaches. According to the Washington State Department of Wildlife, more than 85% of wildlife inhabit riparian areas at some time during their life cycle. And all but the most drought hardy will stop by for some moisture now and then. Streams are indeed important not only for fish, but for all other animals as well.

The word "wildlife" usually refers to critters belonging to four of the five major groups of **vertebrates** (animals with backbones). These four groups include the mammals, the reptiles, the amphibians, and the birds. The fifth major group of vertebrates is fish. One feature that distinguishes fish from wildlife is the fact that fish breathe dissolved oxygen through gills under the water. Generally, animals in the four other vertebrate groups breathe atmospheric oxygen with lungs for at least part of their lives, if not all.

Although less confined to the water than fish, many species of wildlife live much of their lives in aquatic environments. **Amphibians**, which include frogs, toads, and salamanders, are the most water dependent of the non-fish vertebrates. Most lay their eggs in water, and many larval forms actually breathe dissolved oxygen with gills. **Reptiles**, which include snakes, lizards, and turtles, lay their eggs on land, even the species of turtles that spend most of their lives in the water.

While there are many aquatic species of birds and mammals (ducks, gulls, whales, seals, etc), the majority of birds and mammals are terrestrial. **Birds** are the only vertebrate group bearing feathers, and **mammals** are distinguished by the fact that they give birth to live young and nurse them with milk. (There are some members of the other vertebrate groups which give birth to live young, but this trait is the exception rather than the rule for the primarily egg-laying amphibians, reptiles, and birds).

We strongly recommend you don't get caught in the trap of limiting your

WHITEMORE©

wildlife inventory exclusively to the riparian zone. Kevin McGarigal and William C. McComb of the Oregon State University Forest Science Department conducted a study which compared mammals and **herps** (reptiles and amphibians) collected in upland and riparian areas.

These scientist concluded that wildlife managers should consider both upland and riparian habitats if they are interested in maintaining regional biodiversity and providing habitat for a diverse assemblage of wildlife species. They feel that a strong focus on riparian areas with less regard for uplands is inadequate.

Thus, as you use the methods in this book to survey the wildlife in the 500 x 500 square foot area along and adjacent to your stream reach, don't forget about the importance of the rest of your stream's watershed in providing wildlife habitat. The size of the upland area required for wildlife depends on the species. Some species may not require more than the 500 x 500 square foot area that you survey, while many others will.

Vegetation

To complete a stream reach survey, it is important to have a fundamental understanding of vegetation in the **riparian zone**. The riparian zone is the area of land adjacent to flowing water. It is an unique area, containing elements of both aquatic and terrestrial ecosystems which mutually influence each other. The land influences the aquatic ecosystem by contributing organic debris, nutrients, and sediments to the water. The water influences the terrestrial ecosystem by saturating soils, elevating the water table, flooding the surface, and depositing sediments.

Because of the effects of water on the riparian zone, streamside areas typically consist of plants that are water dependent, often referred to as **wetland plants**. Plants that are not water dependent are generally categorized as **upland plants**. However, nature is never that clear-cut. The transition from riparian to upland is more like a continuum of vegetation that changes as the water content of the soil changes.

Obligate wetland plants, such as cattails, skunk cabbage, and some species of willows, grow only in **hydric** soils (soils saturated with water). **Facultative** wetland plants, such as red alder, western red cedar, and stinging nettles, occur in drier soils that are periodically saturated with water; they are found in both wetland and upland areas. Upland plants, on the other hand, will not survive in hydric soils.

Vegetation may also vary with different soil types, textures, erodibility, and the steepness and exposure (north-facing, south-facing, etc.) of the slope it is growing on. Contact your local Soil Conservation Service for information about soils and vegetation in your area. For information on wetland plants, start with *Adopting A Wetland: A Northwest Guide*. It has a great bibliography on wetland publications.

The Importance of Riparian Vegetation

The figures on page 61 illustrate the difference between a healthy and unhealthy riparian area.

The positive role riparian vegetation plays in the health of aquatic and terrestrial system is enormous:

- Vegetated areas help augment surface flows during dry times because they absorb rain and moisture, storing it in the watershed.
- Vegetated areas help prevent flooding and pollution because their ability to absorb rain and moisture reduces and filters runoff.
- Trees provide an overhead shade canopy that keeps stream temperatures cool.
- Logs, root wads, low-hanging branches, and other streamside vegetation that hangs over the water provide protective cover for fish.

figure 4.1 Healthy Riparian Habitat

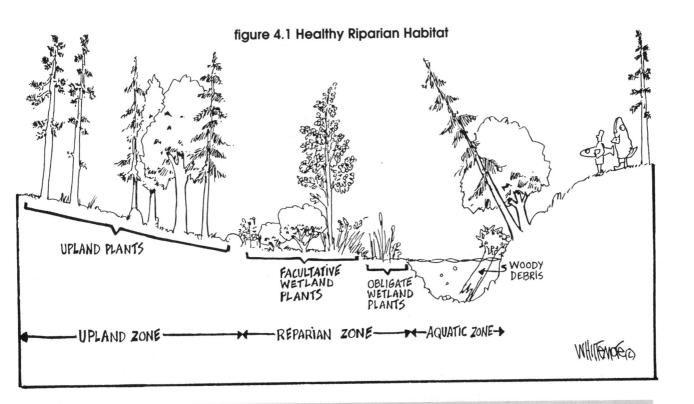

UPLAND PLANTS

FACULTATIVE WETLAND PLANTS

OBLIGATE WETLAND PLANTS

WOODY DEBRIS

←——UPLAND ZONE——→ ←—REPARIAN ZONE——→ ←—AQUATIC ZONE—→

WHITTEMORE ©

A Healthy Riparian Zone

good shade, cool water
abundant woody and organic debris in stream
abundant vegetaion and roots to protect and
   stabilize banks
gravelly, narrow, deep channel
good fish and wildlife habitat
good water quality
high forage production
high water table and increased storage
   capacity
higher late summer stream flows

An Unhealthy Riparian Zone

little shade, warm water
lack of woody and organic debris in stream
little vegetation and roots to protect and
   stabilize banks
silty, wide, shallow channel
poor fish and wildlife habitat
poor water quality
low forage production
low water table and decreased storage
   capacity
reduced late summer stream flows

figure 4.2 Unhealthy Riparian Habitat

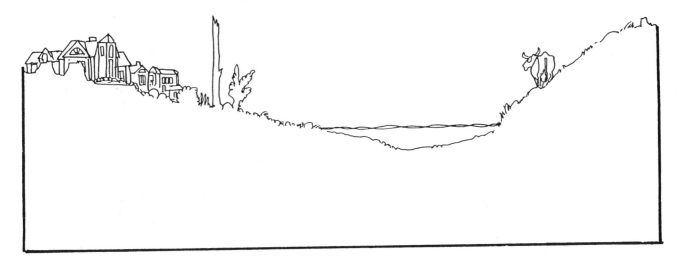

- Trees contribute large woody debris to streams, which creates varied habitats such as riffles and pools, and controls the flow of sediment, keeping it out of spawning gravels.
- Vegetation provides a food source for aquatic invertebates by dropping leaf and other organic material into streams.
- Plant roots along the stream stabilize stream banks and prevent erosion.
- Riparian vegetation provides important habitat for many species of birds and other wildlife.

Stream Channel Gradient, Cross Section, and Sinuosity

The **gradient** of a stream is the change in its elevation over distance, or, in other words, the steepness of the stream. Gradient plays a major role in determining the velocity of stream water, which in turn determines the erosive force of a stream.

Recall the section entitled "The River Continuum" in Chapter 1 that described the physical and biological nature of a stream channel as it progresses downstream. It is the erosive force of water as it flows through watershed hillslopes which shapes the physical character of stream channels and creates the patterns described by the river continuum.

In general, high gradient streams occur on steep, rocky mountainsides. These streams tend to be relatively straight, using their erosive energy to cut a **"V" shaped** valley into the hillslope as well as a "V" shaped channel shape downward into the rock. This channel shape is referred to as the **cross section shape or channel profile** of the stream.

The high velocity of steep gradient streams makes them capable of moving large sizes of stream bottom materials. Their

stream bottoms tend to consist of boulders and larger cobbles. Much of the smaller, lighter materials remain suspended; the current carries them further downstream. A river or stream acts like a continuous series of sieves, with finer and finer stream bottom materials deposited in each reach as the flow moves downstream from steeper to flatter reaches.

Thus as the gradient and velocity diminish, smaller cobbles, gravel, and sand make up the stream bottom. Because the smaller sized substrate is less resistant than boulders and bedrock, the force of the water is able to cut bends in the stream banks. The stream channel takes on a more meandering path.

On the outside bends, the current is strongest, eroding sand and gravel from the bank and cutting down into the channel bottom to form pools. The stream then carries eroded materials downstream, depositing them on the inside bends where the current is weaker. In these **depositional** or **accretion zones**, you will often find gravel or sand bars.

Over time, stream systems in moderate to low gradient areas tend to take on a characteristic pattern with meanders that are five to seven

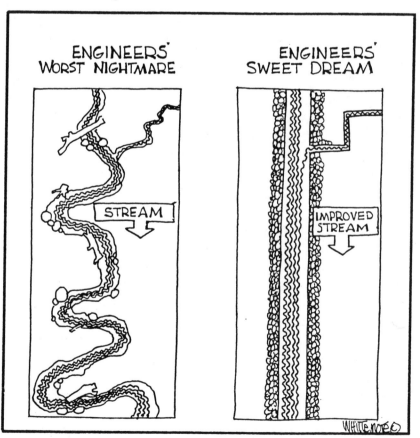

ENGINEERS' WORST NIGHTMARE

ENGINEERS' SWEET DREAM

STREAM

IMPROVED STREAM

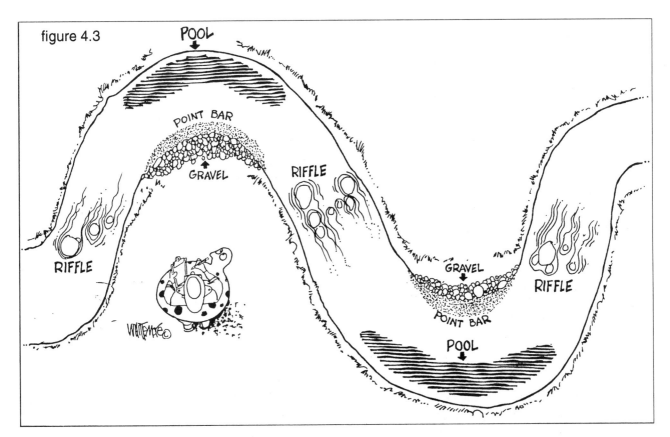

figure 4.3

stream widths apart. The **thalweg**, or path of maximum depth of the stream, follows the meandering pattern, back and forth across the channel. Streams in these moderate to low gradient reaches are able to cut wider, **"U" shaped** valleys through their more gentle hillslopes and "U" shaped cross section profiles through their less resistant substrate.

Some moderate to low gradient systems that cut through fairly broad floodplains will take on a **braided** appearance if they experience greatly fluctuating flows. A braided stream meanders inside its wide channel, separating into many branches, which merge and part again, flowing around gravel deposit islands. Between floods, vegetation may stablize these interchannel islands.

In low gradient areas, the slow moving water allows fine suspended particles to be deposited. Thus these lower reaches generally have sandy or silty bottoms. They tend to meander in wide loops through a flat, fertile flood plain, and the stream profile widens out into a flat "⌣" shape.

Sinuosity is the degree to which a stream channel has a meandering or straight pattern. Except for high gradient areas where streams flow rapidly through boulders or bedrock, stream channels are seldom straight unless people have

channelized them, forcing them to flow in a straight path by physically altering the channel.

When a stream is left to its natural meandering pattern, the erosive force of the flow is dissipated as the water travels a longer distance through the bends and curves. This results in a more stable system than a channelized stream whose erosive energy simply builds as it is funneled straight downstream.

A watershed drainage network continuously attempts to establish a balance between the shape of its stream channels and the amount and force of water running off from its hillslopes. In mature, stable stream reaches, the amount of sediment and water that enters the reach equals the amount that leaves it. This balance point is referred to as a **dynamic equilibrium**. To maintain this state, the stream and its channel will change in response to changing conditions in the watershed.

Thus the physical character of the channel offers clues to the health of the stream and its watershed. If a land use alters the amount of water and sediment that runs off into a stream, the stream and its channel will try to adjust. Depending on the resistance of the substrate, the addition of water and sediment may erode the stream bed in different ways.

SALMON PERTURBATIONS

LEAVING HIS POOL WOULD START A SERIES OF EVENTS OVER WHICH HE HAD NO CONTROL. WITH AN IMPENDING SENSE OF DOOM HE LET HIMSELF SLIP INTO THE MAINSTREAM.

A diverse mix of mature riparian vegetation growing on the banks of a stream indicates that the banks are relatively stable. Although the product of erosive forces, bank undercuts can also be stable if they are well vegetated. Stable bank undercuts provide excellent shaded, protective habitat for fish.

On the other hand, a stream bank with little or no vegetation usually indicates that excessive bank erosion is occuring. Signs of erosion include **rills** and **gulleys** cut by runoff flowing down the banks into the stream channel. **Sloughing** or collapsing banks are another obvious sign of erosion. Unstable bank undercuts could threaten rather than provide good fish habitat, as they are highly prone to sloughing.

Armoring banks with rock or concrete may be effective in remedying emergency erosion problems. However, artificial bank stabilization tends to deflect and concentrate the force of stream flow down-stream to unprotected areas. Thus, forcing a stream into an unnatural human-preferred course may accelerate erosion downstream. It also limits the stream's ability to create a natural diversity of habitats. The best form of bank protection is to leave riparian vegetation intact and keep stability-disturbing activities away from stream banks.

In areas with deep **alluvial** (flood derived) soils, increased runoff can downcut the stream channel and scour out the stream bottom. In areas where rock bottoms resist downcutting, the result may be a widened stream channel and a stream bottom covered up with sediment.

Stream Banks

Integral with the shape of the stream channel is the condition of the stream banks in offering clues about the health of a stream. As mentioned above, a certain amount of scouring or erosion generally occurs on the outside bend of stream meanders, where water velocity is greatest.

This is where you will find **bank undercuts**, areas where the bank overhangs the stream. Where water flows with less velocity at the inside bend of meanders, you often find accretion areas where gravel, cobbles, and sediment are deposited. In general, stream banks are relatively steep on outside bends and shallow at inside bends. The shape of this type of channel is shown at the top of the next page.

Instream Habitat

There are various features within and along stream channels that provide important habitat for stream organisms:

- **Riffles** are shallow areas of fast moving white water. The cobble/gravel bottoms and well-oxygenated water of riffles provide optimal conditions for many underwater insects and spawning salmonids.

figure 4.4

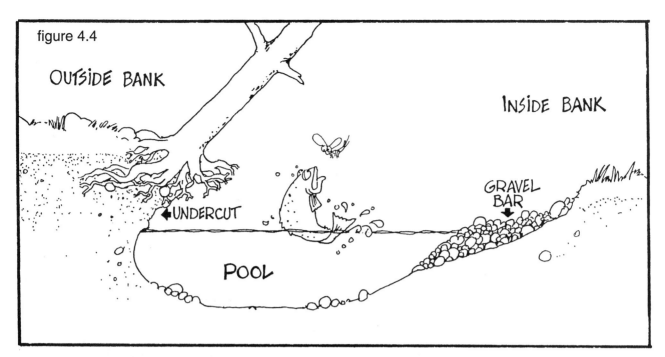

A cross sectional view of a meander bed

- **Pools** are deeper areas where water slows down. They provide hiding places and refuges for fish to rest from the current.
- **Runs** are fast flowing areas but not shallow enough that the substrate creates white water riffles.
- **Glides** are relatively slow moving areas, but still faster than water moving through pools.

A diverse habitat situation is the best for salmonids. A pool to riffle ratio of 1 to 1 is considered by most biologists to be the ideal. Other factors will affect a stream's fish productivity, however, and you may find very productive streams with a much lower or higher pool to riffle ratio. Runs and glides also provide important food and habitat for fish, but without pools and riffles, fish habitat is significantly less varied and productive.

figure 4.5

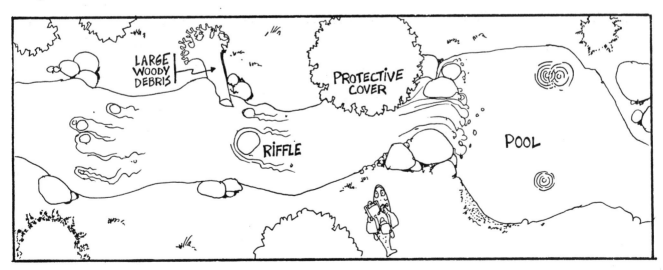

figure 4.6

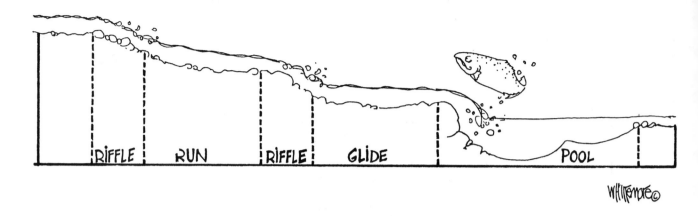

RIFFLE   RUN   RIFFLE   GLIDE   POOL

Most of the obvious pools and riffles will occur in streams with moderate gradients and cobble/gravel bottoms. In steeper, boulder-laden headwater streams, shallow riffles and pools are less common than steeper **cascades** and **plunge pools** or **step pools**. As the gradient flattens out in the lower reaches of a stream system and stream flows slow down, you may find that glides are most common.

The **side channels, streamside wetlands,** and **sloughs** that form as streams progress downstream to their mouths provide excellent rearing areas for fish that may spawn in the higher gradient areas of the stream. These areas also provide important habitat for a variety of other wildlife species.

Woody debris, such as fallen trees and branches, enters the stream from the adjacent riparian zone. It creates a diversity of habitat types, including pools that often form downstream from fallen logs.

Woody debris also traps sediment and smaller **organic debris**, such as leaves and twigs. It prevents spawning areas from being inundated with large sediment loads and it retains organic matter long enough for it to break down and become more appetizing to aquatic invertebrates.

Woody debris also creates protective cover for fish. When logs and rootwads hang over the stream, they create excellent hiding places for fish to seek refuge from both terrestrial and aquatic predators. Other forms of cover include riparian vegetation that hangs down close to the water, aquatic vegetation, undercut banks, and boulders.

Human Alterations to the Stream Channel

Many human-made objects and structures have been placed in and along streams to remove water, to divert the flow, and to dispose of storm water. They may impact streams in many ways, such as significantly reducing or altering flows and contributing effluents.

As mentioned on page 63, artificial bank protection such as rip-rap, while solving immediate erosion problems, often creates other problems downstream. Channelizing a stream disrupts its ability to stabilize the force of its flow and significantly reduces habitat structure for fish and other aquatic critters.

When we build roads and buildings over and near streams, streams end up underground, with their water channelled through **culverts**. Culverts can be major barriers to fish passage if undersized or not installed correctly. They can also cause scouring of the stream channel and be a source of pollutants. Of course a bridge over the stream that does not affect the stream channel will prevent these problems.

Some artificial structures do benefit streams. **Detention ponds** are built downstream from areas developed with impervious surfaces. They slow and retain storm water, allowing sediments to settle out before the water is released into streams.

Field Procedure: Stream Reach Map

EQUIPMENT NEEDED

folding map board (see "Making a Folding Map Board," below)

pencils, water soluble marking pens

tape measures, one short, 30' - 50'; two long, 100' - 200'

compass

stakes or rebar

forestry tape

waterproof boots or waders

WHAT?

This procedure helps you document the location and path of the stream channel. As you do so, you'll also document various features associated with the channel and the area on either side of the stream. Once you complete your base map, you can use it to document information gatheed in your stream reach survey.

WHERE?

The base map will cover the 500 x 500 square foot area that includes your 500 foot- long stream reach, as well as the area extending 250 feet from either side of the stream, along the length of the reach. Make a base map for each stream reach that you monitor.

WHEN?

Map your stream reaches each time you survey them. We recommend you conduct stream reach surveys four times a year: once in the spring, summer, fall, and winter, to detect seasonal changes that occur in your stream.

You may only need to create a base map once for each reach, the first time you survey the reach. When you come back for the second survey, make a duplicate of the original base map and document on the duplicate any changes that have occurred.

Continue this process each time you return to survey the reach, documenting current conditions on the most recent version of your base map.

HOW?

Scale

Use graph paper to sketch your stream reach map. This will help you determine and stick to a **scale** so that you can record your observations precisely. A scale tells you how many square feet or meters on the ground each square (or inch) on your graph paper represents.

Make sure you indicate the scale you choose on your map. An 8.5" x 11" sheet of graph paper with 4 squares to the inch can provide a map of a 500' x 500' area if you use a scale of one square equals 20 feet (or one inch equals 80 feet). However, this scale will not provide much detail of your stream.

We recommend that you use flip chart sized graph paper (about 25" x 30") that is divided into one inch squares. If you use a scale of 1 square equals 10 feet (one inch equals 10 feet) and piece together four flip chart sheets, you can create a very detailed map of the entire 500' x 500' area.

Of course, this large a piece of graph paper can be unwieldy to use in the field. What works well is to mount the flip chart sized graph paper onto corrugated cardboard or foam core board. Instructions on how to make this handy tool follow.

Making a Folding Map Boar

Measure the size of your graph paper. Cut two pieces of foam core board or cardboard to match that size (we have been using the same foam core map board for three years; cardboard won't last as long). Attach the two boards with duct tape so they open and close like a book. Mount two pieces of flip chart graph paper on your "book."

Using the scale mentioned above (1"=10'), this folding map board will cover one half of your 500' x 500' area. To map the entire area, make a second map board.

To weatherproof your folding map boards and make them reusable, cover them with clear contact paper or acetate. When out in the field, use a water soluble marking pen to draw your map. When you return from the field, copy your map onto paper, wipe off the field copy, and your map boards will be ready for the next field mapping exercise.

Mapping Methods

Depending on the nature of your stream reach, you can use one of two different mapping methods described below. There are benefits and shortcomings to both approaches.

The first method uses a 500 foot long baseline that parallels your stream. You map the stream channel by walking along lateral (perpendicular) transects and recording the distance of the channel from the baseline. In the second approach, you map the stream by walking the channel itself, and lateral transects are used only to survey the surrounding area.

The first method is likely to yield more accurate results because distances are recorded from a straight baseline. However, it may be extremely difficult to carry out in areas where steep hillsides surround the stream.

The second method can be used in steep areas, but may be slightly less accurate and your lateral transects will likely overlap each other. You'll discover, however, that it gives you a more accurate picture of the sinuosity of your stream. (See section on sinuosity, page 62.)

No matter which method you use, make sure you mark the beginning of your reach with permanent visible markers such as stakes or brightly colored forestry tape to make it easy to find during follow-up trips.

Method I: For Streams in Flat Valleys

1. Start at the designated beginning of your stream reach. A short distance from the channel, measure off a straight baseline parallel to your stream (eventually you will extend this for 500 feet). If there is a fairly straight trail or road that parallels your stream, you can use it to set up your baseline. This will make your mapping easier.

2. Draw your baseline on your map as a vertical line. Use a compass to determine the magnetic direction of your baseline and record this on your map. If you use a trail or road as your baseline and it changes direction, make adjustments on your map accordingly.

Note: Since you are measuring 500 feet along a straight baseline, not along your meandering stream, the actual length of the stream that you map will be greater than 500' unless the section of stream is very straight.

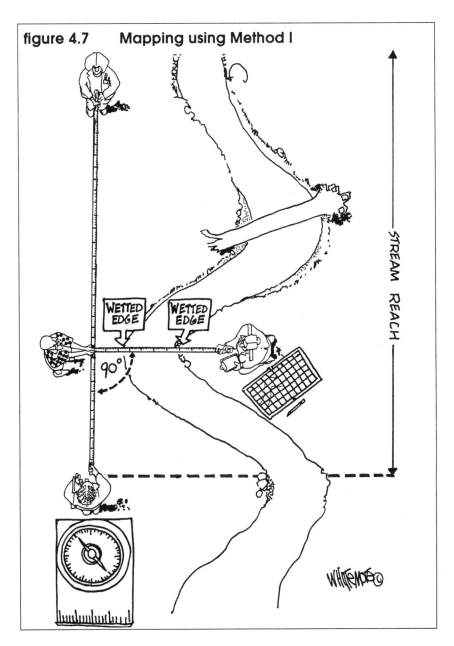

figure 4.7   Mapping using Method I

3. Extend lateral transect lines from the baseline to the stream. These transects should be perpendicular (at 90° angles) to the baseline and occur at regular intervals along the 500-foot long stretch. We recommend setting up a minimum of ten transect lines at 50' intervals. To be more precise and accurate, add more transect lines. The more transect lines you use, the more accurate your map will be. Use your compass to make sure the lateral transect lines are indeed perpendicular to your baseline. Use stakes or forestry tape to mark the location of your transect lines along the baseline.

4. Along each transect, measure the distance from your baseline to each **wetted edge** of the stream (the points where the water surface hits the banks). Mark an "X" on your map at the locations where each transect crosses a wetted edge. When you have done this for all transects, you'll be able to draw your stream by connecting the "X's."

5. Sketch as much information as you can about the stream at the area where the transect crosses the stream. Include observations about all features that will be part of your stream reach survey, such as fish and wildlife sightings, vegetation, stream banks, instream habitat, and human alterations.

6. If your stream ends up straying more than 20-30 feet from the baseline, add another marker in the field at each point where the transect line crosses the stream. These markers will be important guides when you go back and survey your stream channel in more detail.

7. Extend each transect line from the baseline to a distance of 250 feet on either side of the stream. Again, make sure each transect is perpendicular to the

baseline. Mark the boundaries of your 500' x 500' area in the field with stakes or forestry tape. Draw the boundaries on your map, and document any landmarks and/or vegetation that occur at the end of each transect line.

Note: Since you are measuring your transect lines 250 feet out from the stream, the external boundaries of your 500' x 500' area will vary unless your stream is very straight. See sample of completed base map using Method I, figure 4.8.

8. There are three parameters in the stream reach survey that are observed in the area surrounding the stream: riparian vegetation, wildlife, and land uses. Specific instructions on how to survey and map these features are provided in the appropriate field procedure sections that follow. If you would like to survey and map these features at a later time, you can place markers along the transect lines, at regular intervals (50' or less), from the baseline to the outer boundaries of your 500 x 500 square foot area. Then you can use these markers as guides when you return to conduct your surveys.

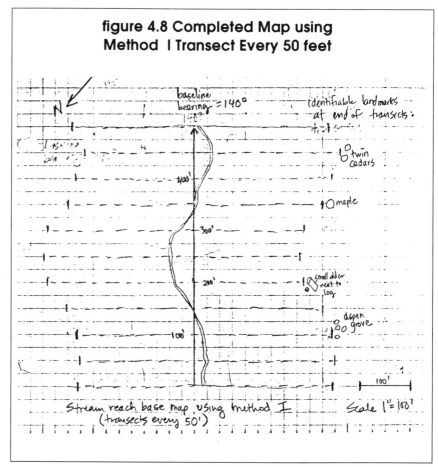

figure 4.8 Completed Map using Method I Transect Every 50 feet

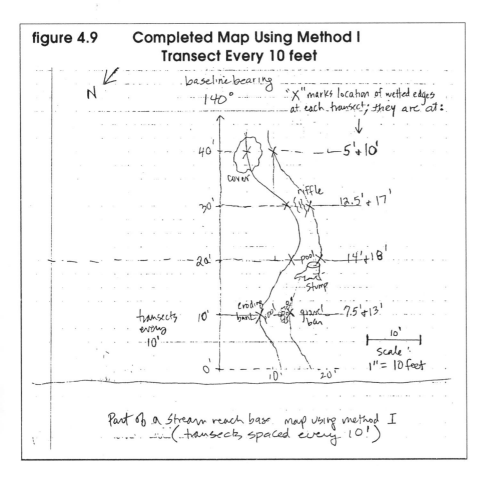

**figure 4.9   Completed Map Using Method I
Transect Every 10 feet**

N

baseline bearing
140°

"X" marks location of wetted edges
at each transect; they are at:

40' — 5' + 10'

cover

riffle
30' — 12.5' + 17'

20' — pool — 14' + 18'

stump

transects
every
10'

eroding
bank
10' — pool — gravel
bar — 7.5' + 13'

10'
scale:
1" = 10 feet

0'   10'   20'

Part of a stream reach base map using method I
(transects spaced every 10'!)

Method II:
For Streams in Steep Valleys

1. Start in the middle of your stream (feet in the water) at the beginning of your designated reach. Extend a baseline, in the middle of the stream, from the beginning of the reach to the first bend in the stream. Use your compass to determine your direction to that point, and measure the distance from the starting point to the bend. Draw the baseline and record the compass direction on your map.

2. At regular intervals, set up lateral transect lines at 90° angles to the stream/ baseline. We recommend a minimum of 11 transect lines at 50' intervals. Remember, the more transect lines you use, the more accurate your map will be. Use your compass to make sure the transects are indeed perpendicular. Use stakes or forestry tape to mark the location of your transect lines along one bank of the stream.

figure 4.10 Mapping Method II

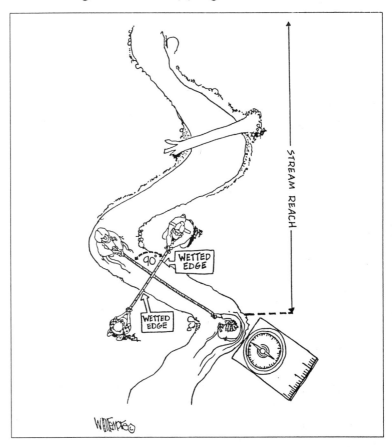

STREAM REACH

90°

WETTED
EDGE

WETTED
EDGE

3. At each transect line, measure the width of the stream, from wetted edge to wetted edge. Mark an "X" on your map at the locations where each transect crosses a wetted edge. When you have done this for all transects, you'll be able to draw your stream by connecting the "X's."

4. Sketch on your map as much information as you can about the stream at the area where each transect crosses. Include observations about features that will be part of your stream reach survey, such as fish and wildlife sightings, vegetation, stream banks, instream habitat, and human alterations.

5. Extend each transect line from the stream/baseline to a distance of 250 feet on either side of the stream. Again, make sure each transect is perpendicular to the baseline. Mark the boundaries of your 500' x 500' area in the field with stakes or forestry tape. Draw the boundaries on your map and document any landmarks and/or vegetation that occurs at the end of each transect line.

Note: Since your baseline follows the stream, your transect lines will overlap unless your stream is very straight. Also, the external boundaries of your 500 x 500 foot area will not form a neat square.

6. Locate the next bend in your stream and repeat the process. You will continue this exercise through the stream reach, ending at a total distance of 500 feet.

7. There are three parameters in the stream reach survey that are observed in the area surrounding the stream: riparian vegetation, wildlife, and land uses. Specific instructions on how to survey and map these features are provided in the appropriate field procedure sections that follow. If you would like to survey and map these features at a later time, you can place markers along the transect lines, at regular intervals, from the baseline to the outer boundaries of your 500 x 500 square foot area. Then you can use these markers as guides when you return to conduct your surveys.

NOW WHAT?

Use your base map to document the more detailed information that you collect during your stream reach survey.

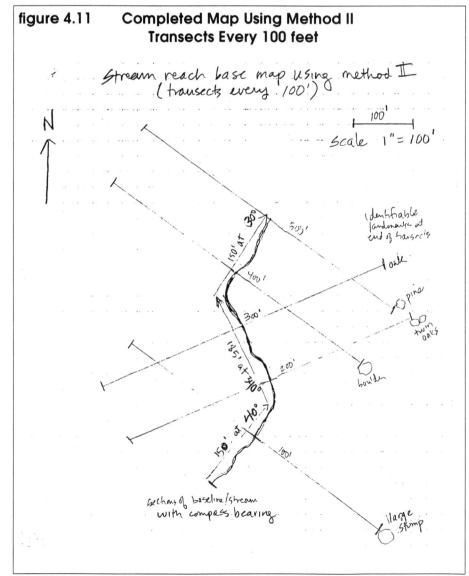

figure 4.11   Completed Map Using Method II Transects Every 100 feet

FIELD PROCEDURE: STREAM REACH SURVEY

EQUIPMENT NEEDED

your stream reach base map
Other map, e.g., USGS topos, aerial photos, etc.
Data Sheet 2, "Stream Reach Survey"
clipboard
pencils
tape measures, one short, 30' - 50'; two long,
    100' - 200'
hip chain (optional)
compass
stakes or rebar
forestry tape
waterproof boots or waders
camera & film
plant, fish, and wildlife field guides
polarized sunglasses
strong string, at least 100'
carpenter's level
pole and yard stick, or stadia rod
    (see figure 5-2, page 97 in Chapter 5)

WHAT?

The parameters covered in the stream reach
    survey include:
- fish
- wildlife
- vegetation
- overhead canopy
- stream channel gradient
- channel and valley cross-section shape
- channel sinuosity
- stream banks
- instream habitat
- human alterations to the stream channel
- land uses adjacent to the stream reach

WHERE?

You'll conduct your stream reach survey along the length of your 500' long stream reach, as well as in the area extending 250' from either side of the stream. This should be the 500' x 500' area covered by your stream reach base map.

WHEN?

For each stream reach that you are monitoring, we recommend you conduct a survey **four times a year**, once in the spring, summer, fall, and winter, to detect seasonal changes that occur in your stream. If you do not have the resources or time to survey quarterly, then survey once during the low flow time of the year, and once after the peak runoff season. In most areas, this will be late summer and early spring.

HOW?

The first step in a stream reach survey is to document the exact location of the beginning of your stream reach for the record. See the section entitled "Documenting Locations" on the following page for methods on how to do so. Make sure you have also marked the beginning and end of your reach in the field with some kind of permanent markers such as stakes or forestry tape if you haven't done so already. These markers will serve as a guide as you survey your reach, and will help you find it again when you return for future surveys.

After you have documented the location of your stream reach, you are ready to survey the parameters listed in the "What" section. Detailed field procedures for each parameter are found on pages 78-91. You have already documented some information for many of these parameters on your stream reach base map. During your stream reach survey, you will add more detail to your map as well as record all information on Data Sheet 2, "Stream Reach Survey." You'll find that some information is difficult to illustrate on a map and thus more appropriate to record on a data form.

Photographs are also extremely valuable tools for documenting information on your stream reach. A series of photos taken over a period of time is an effective way to tell the story of how your stream changes. They can also help you find your way back to the same spots when you return to your stream reach for future surveys. If you decide to use photographs to complement your maps and data sheets, record the location of each photograph and the compass direction in which it was taken. This way, you can take repeat photographs that provide an accurate comparison of changes.

DOCUMENTING LOCATIONS

This section provides three different methods for documenting the exact location of the beginning of your stream reach. You can also use these methods later to document the locations of the specific sites within your reach that you select for further study.

Driving and Hiking Directions

Record clear driving and hiking directions to your stream reach and study sites. Be very explicit about the locations of landmarks such as bridges, buildings, large trees, your own field markers, or other features. Use compass directions if necessary. Attach appropriate maps to ensure that anyone can find the same locations.

River Miles

The location of a site in river miles is the distance of that site from the site to the mouth of the stream or river. If a road follows the path of your stream, you can determine the river mile location of the beginning of your stream reach and your study sites by clocking the distances with your car's odometer.

If you don't have this luxury (which is nothing to hope for, as a road will impact your stream), use a map to estimate river miles. Aerial photos and large scale (highly detailed) topographic maps are the best tools for this exercise, because they have details that can help you find specific sites on them. Lay a string along the path of your stream from its mouth to the beginning of your stream reach. Then use the scale on the map to determine the actual distance this length of string represents. Use the same procedure to estimate the river mile location of your study sites.

Section, Township, and Range

Even by learning the section, township, and range your stream is in, you will not know its exact location. However, it is important to have this information so you can find your reach on any map. For more information on this type of locator, refer to page 22 in Chapter 2.

Latitude and Longitude

A few thousand years ago, Greek and Egyptian mathematicians and geographers divided the surface of the earth with two sets of imaginary lines. These lines, known as latitude and longitude, were used for navigation and to pinpoint exact locations on the earth. They are still used today.

Lines of **latitude** are imaginary east-west lines around the circumference of the earth which are parallel to the equator. These lines indicate a direction in degrees north or south of the equator, from 0° at the equator, to 90° at the poles. Each **degree** is divided into 60 minutes (60') and each **minute** into 60 seconds (60"). At the equator, one minute of latitude equals one nautical mile.

Lines of **longitude** (or meridians) are north-south lines which extend from north pole to south pole at right angles to lines of latitude. The starting point or 0° longitude is the **Prime Meridian** which runs through Greenwich, England. Longitude is measured in degrees east and west from the Prime Meridian to halfway around the world at the 180° meridian in the Pacific Ocean, called the **International Date Line**.

figure 4.12

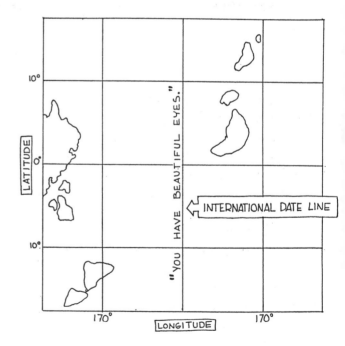

figure 4.13

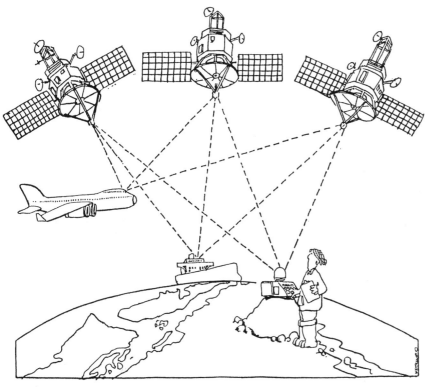

Any point on earth can be identified by its north-south latitude and east-west longitude. The latitude and longitude values are known as **coordinates**. You can use coordinates of latitude and longitude to indicate the location on earth of the beginning of your stream reach or your study sites.

The benefit to finding the coordinates of your reach and sites is that more and more agencies and organizations are using a computerized **geographical information system (GIS)** for all kinds of work that involves maps and other geographical information. GIS uses databases in which all information is stored as coordinates of latitude and longitude. If you know the coordinates of your stream reach and study sites, anyone with a GIS system, from an agency employee to an interested citizen from another country, will be able to find them.

You can calculate the latitude and longitude of a specific site mathematically if you have a USGS topographic map and can locate the site on that map. This mathematical approach is provided in the "Worksheet for Calculating Latitude and Longitude" on page 76.

Another tool that can be used for determining coordinates is an aeronautical chart. These are maps used by pilots, showing lines of latitude and longitude. If you can find the location of your stream reach and study sites on an aeronautical chart, you will be able to determine their coordinates using a ruler.

The high-tech tool for determining coordinates is a **global positioning system (GPS)**. A GPS is a hand-held electronic device that uses satellites to determine the latitude and longitude of the location where the person holding the device is standing. You can take a GPS out in the field to determine the coordinates of different sites along your stream. The accuracy of these systems is improving, and electronic beams from up to five satellites at a time can pinpoint your position on the globe to within 15 feet or better.

You can buy these systems for costs ranging from $150 to $3000, or you may be able to borrow a GPS from a local yacht club or airport, as they are commonly used for navigation. More and more government agencies are using these systems as well----another possible source of a "loaner."

. . . with apologies to A.A. Milne and Ernest H. Shepard.

TABLE 5 WORKSHEET FOR CALCULATING LONGITUDE AND LATITUDE

(You will need your 7.5 minute USGS map, a metric ruler and a calculator)

LONGITUDE
1. Look at the right hand corner (upper or lower) under the map name
   (last digits of number) to find the width scale (longitude) of the map:
   If "7.5 Minute Series," enter 450
   If "15 Minute Series," enter 900
   If "7.5x15 Minute Series," enter 900   _____

2. Record the east to west width of your map, (excluding white border)   _____cm

3. Divide #1 by #2 (Round off to the nearest whole number) (Longitude factor)   _____ sec/cm

4. Enter the longitude for the edge closest to your site
   (Longitude is at the corners of your map and will be over 100°)   _____° _____' _____"

5. Record the distance, in cm, from the site to the map edge   _____ cm

6. Multiply #5 by #3 (round off to the nearest whole number).   _____ sec

7. Convert #6 to minutes and seconds by dividing by 60. (Don't use a
   calculator) After division, your whole number is minutes and the
   remainder is the number of seconds. For example: 150/ 60 = 2 R30
                              150 = 2 min 30 sec. (2' 30")   _____ min _____ sec

8. If the map's closest edge is east of the site, add #4 to #7
   If the closest edge is west, subtract #7 from #4
   THE ANSWER FOR #8 IS THE LONGITUDE OF THE SITE

   _____° _____' _____"
   LONGITUDE

LATITUDE
9. Look at the right hand corner (upper or lower) under the map name
   (first digit of number) to find the height scale (latitude) of the map.
   If "7.5 Minute Series," enter 450
   If "15 Minute Series," enter 900
   If "7.5x15 Minute Series," enter 450   _____

10. Record the length of your map from north to south   _____cm

11. Divide #9 by #10 (round off to the nearest whole number). (Latitude factor)   _____sec

12. Enter the latitude for the map edge closest to your site (north-south)   _____° _____' _____"

13. Measure (in centimeters) from the site to the map edge   _____cm

14. Multiply #13 by #11 (to the nearest whole)   _____

15. Convert #14 to seconds (explanation from #7)   _____' _____"

16. (a) If map's closest edge is south of site, add #15 to #12
    (b) If map's closest edge is north of site, subtract #15 from #12

THE ANSWER FOR #16 IS THE LATITUDE OF THE SITE
   (Make sure the numbers are between the longitudes and latitudes
   that surround your site.)

   _____° _____' _____"
   LATITUDE

Adapted with permission from the U.S. Environmental Protection Agency's *Streamwalk*

A furry Streamkeeper makes
a legend for a fish survey
sketch map.

FIELD PROCEDURE: FISH

Note: The remaining text in this chapter corresponds with the necessary data to be recorded on Data Sheet 2: Stream Reach Survey.

Surveying the characteristics of a stream gives you a good idea of the potential for the stream to provide habitat for fish. Surveying the fish populations themselves gives you direct information on what is present. Because of their elusiveness and mobility, however, fish are difficult to survey. Moreover, stream habitats and fish habitat needs change with the seasons. If you plan to implement any of the fish survey techniques discussed below, we recommend you contact your state's fisheries or wildlife agency for advice and assistance. It is also important to check with them to make sure you will not be disrupting any fish migration or spawning activity.

Use Data Sheet 2 to record the presence of fish, and if possible, the numbers and species observed. If you see any dead fish, note any evidence that may help you determine if they are spawned out carcasses versus victims of some other cause of death. Mark the location of your observations on your stream reach map.

VISUAL METHODS

Simple visual methods can be used to note the presence or absence of fish. Identifying fish by species using one of the following visual approaches is a challenge. You cannot put the "fish in hand" and compare them to a fish key. However, if you investigate your watershed prior to starting the stream reach survey, and consult your local fisheries biologist, you will have a pretty good idea of what species of fish you should find in your stream at a particular time of the year.

The visual approaches in Methods 1-3 below will only work in relatively clear water. For the spawning, carcass, and redd survey methods 4 and 5, make sure you ask your fisheries biologists what times of the year you're most likely to have success.

1. The easiest fish survey method is the **sit and wait** approach. It only requires a pair of polarized sunglasses and a bit of patience. Find likely "fishing holes," such as pools or undercut banks, make yourself comfortable, and sit and wait. Polarized sunglasses help significantly to cut the glare of sunlight on the water.

2. Another simple visual technique is to hold a **glass bottomed box** over a "fishing hole," stay quiet and still, and watch for any fish that swim by. You can make a "glass bottomed box" by cutting out the bottom of a dishpan or plastic box and glueing in a piece of clear Plexiglas.

3. If you have snorkeling training and gear, you might want to attempt a **snorkel survey**. This literally means getting down in your stream with the fish. We have found that keeping still and

waiting for the fish to come to you yields the best results. Using snorkeling equipment makes it easier to identify the species of fish that you see than if you use method 1 or 2. We suggest that you consult your local fish and wildlife agency before attempting this approach.

4. **Spawning surveys** can be conducted from the stream banks with relative ease because salmon and trout are more visible and less mobile when they are spawning.

5. **Carcass and redd surveys** also provide important information on the status of fish runs. **Redds**, salmon and trout nests, are elongated oval depressions in the streambed that are algae free.

"FISH IN HAND" METHODS

"Fish in hand" methods require prior training and specialized equipment. In addition, these approaches usually require a scientific sampling permit because they involve capturing fish. Contact your local fisheries or wildlife biologists before using these methods.

1. The most common "fish in hand" method biologists use to conduct fish population surveys involves use of an **electroshocker.** This gadget sends a jolt of electricity through the water. The shot of electric current temporarily stuns any fish in the immediate area, causing them to float to the surface.

The good thing about this approach is that you get immediate results. The downside, however, is that if the "juice" is not set precisely, you can kill rather than stun the fish. If you spring a leak in your boots, you can have a "shocking" experience as well.

Not only are electroshockers potentially dangerous, they are expensive as well ($1200 to $1500). Make sure that you are well trained be-

fore you put this technique into practice yourself. Better yet, invite your local fish and wildlife agent to do the survey with you. Not only will you get an expert, but free use of the equipment.

2. **Hand seining** is another common "fish in hand" method. While this approach can be more easily used by Streamkeepers than electro-shocking, it still does require considerable time and skill. For those of you who feel like attempting hand seining we recommend that you follow the procedures developed by the Oregon Department of Fish and Wildlife, reprinted in Appendix A, with permission and minor modifications, from their publication *The Stream Scene, Watershed, Wildlife and People.*

3. **Trapping** fish using a baited trap (cat food works well) is an effective way of catching juvenile fish. Just make sure that you check the trap regularly. "Key" the fish and leave them unharmed. Again consult your local fish and wildlife agency first---a sampling permit may be required.

FIELD PROCEDURE: WILDLIFE

The visual observation approaches recommended below can yield valuable information on the presence and absence of wildlife and their relative abundances in different survey areas. For identification, there are many publications and keys available to help you. Since most wildlife do not appreciate human observers and generally hide from view, be sure to add a track key to your wildlife identification tool box as well.

You may want to bring in an outside partner from your local Audubon Society to help identify birds. Also contact your local fish and wildlife agency; they may ask your Streamkeeper group to "hook into" their wildlife survey programs. Streamkeeper wildlife data can provide a significant addition to local and regional wildlife agency databases.

It is best if your wildlife survey is your first item of business each time you return to your stream reach survey site. This will increase your chances of seeing or hearing wildlife before your presence scares them away.

BIRDS

Birds may be the easiest creatures to spot. Look and listen for them as you walk slowly along your transects from the stream to the outer boundaries of your 500 x 500 square foot area. Take into consideration that different species inhabit different layers of the vegetation, from the dense, shrubby undergrowth to the top of the canopy.

Note the locations where you see and hear birds both spatially and vertically. Record the species that you are able to identify by sight or sound. Also document any evidence you see that birds are present, such as nests, egg shells, feathers, or tracks.

HERPS

Herps (reptiles and amphibians) are more secretive than birds. The best places to look for them are under rocks and logs in damp spots, especially along the stream's wetted edges. Take care not to create too much disturbance on the

Otter and salamander illustrations on these two pages by Sandra Noel, from *Adopting a Steam: A Northwest Handbook*

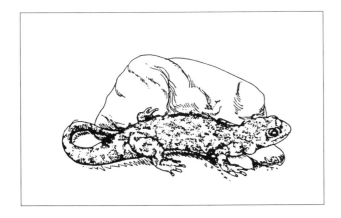

fragile stream banks, and make sure you return all logs and rocks to their original positions.

In the spring, look for amphibian egg masses in slow moving backwater areas, side channels, and other wetlands associated with the stream and in your 500 x 500 square foot area. If you can, take a trip to your stream reach on a spring evening to listen for breeding frogs.

MAMMALS

Mammals may be the most challenging to spot, but they often leave behind evidence of their presence. Note any signs of mammals that you see in addition to actual sightings. Signs may be in the form of nibbled plant matter, scat, bones, fur, tracks, trails, nests, burrows or dens.

RECORDING WHAT YOU FIND

1. Record what you see in the space provided on Data Sheet 2. You can document the species you see, or use broad categories to record your wildlife sightings. For example, birds are categorized into types such as gulls, ducks, sandpipers, passerines (perching birds), raptors, etc. Herps can be easily distinguished as frogs, salamanders, snakes, turtles, etc.

2. You should note significant fish and wildlife sightings on your stream reach map. Estimate the location of your findings or get measurements for greater accuracy. For the detailed approach, use the transects you set up in your mapping exercise. Using a compass in combination with **pacing techniques,** a **hip**

Pacing Techniques

You can map the approximate location of an object using pacing techniques, a reference point, and a compass. Figure out how many of your paces equals a certain distance (e.g., 4 paces = 10 feet). Use a reference point in your 500 square foot area that you can locate both on your map and in the field. (If you are using transects, any given point along a transect line can be a reference point). Record the compass direction from that point to the object, and the number of paces it takes to travel from the point to the object. (Try to use consistently sized paces). Now you have all the information you need to draw the object on your map in its approximate location.

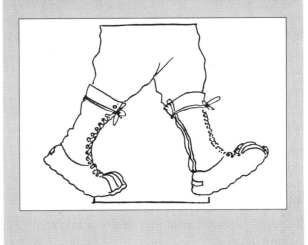

chain, or a tape measure, you can measure the distance and direction a sighting lies from a particular point along a transect line. Then you can easily transpose the information onto your stream reach map.

3. If you happen to find evidence of an endangered, threatened, or sensitive species, make sure you document it with photographs. If you record the location, date, and time that you take any photographs, and have witnesses along, you will have credible data.

figure 4.14

OTHER TECHNIQUES

1. If your time, interest, and patience allows, we recommend the **sit and wait** approach for increasing your wildlife viewing chances. After you become more familiar with your 500 x 500 square foot area, you may be able to identify key locations of potential wildlife habitat. "Stake out" these likely sites, sitting and waiting for stories to unfold.

2. By constructing a **blind** at a key observation point, you may further increase your odds of viewing wildlife. A blind is a structure that allows you to view wildlife without being seen by whatever you are viewing, thus without scaring them away. Blinds usually consist of a wooden wall with "peep" holes.

3. We recommend that Streamkeepers not consider using **trapping techniques** unless there is a very specific and immediate need for the information that you expect to obtain. Generally, trapping is extremely time consuming and potentially injurious to the subject animal. These studies are also very labor intensive, requiring permits and significant advanced planning. If you are interested in more in-depth wildlife surveys, consult local wildlife professionals or your state wildlife agency for assistance.

Using a Hip Chain

A hip chain can be used to measure relatively long distances much more accurately than pacing techniques. A hip chain is a plastic or leather case containing biodegradable string which feeds through a counter.

Attach the end of the string to a known reference point. Set the counter to 0 and use a compass to walk in a straight line to the object you are mapping. The counter will give you the distance from the reference point to the object. Use this distance measurement and your compass direction to locate the object accurately on your map.

Reattach the string and reset the counter for each object that you measure.

FIELD PROCEDURE: VEGETATION

This section outlines methods for surveying the vegetation in your 500 x 500 square foot area. They are fairly simple methods designed to give you a general idea of the types and location of vegetation along and adjacent to your stream reach.

If your objectives require you to collect quantitative data on plants around your stream, we recommend that you contact local experts or appropriate agency personnel in your area who can assist you with designing your study. Who you consult and the study you design will depend on your objectives.

For example, you may want to document the **board feet** or dimensions of trees that are being harvested from a forest around your stream to ensure that a logging operation is in compliance with agency permit regulations. In this case, you would consult a forestry expert.

If you are interested in surveying plant populations that may be relatively rare and deserve special protection, you might want to consult your state's Natural Heritage Program (usually administered by State Departments of Natural Resources, Public Lands, etc.)

No matter what your objective, get to know the vegetation around the stream reach you are surveying and develop a general knowledge of the predominant vegetation in the entire watershed surrounding your stream. Use plant keys and publications available at your local libraries and bookstores to assist in identifying plant species. There may also be a local native plant society with members eager to assist you with plant identification and surveying.

Use the following steps; record your observations on Data Sheet 2.

1. Evaluate the types of vegetation in the overstory (taller trees) in your 500 x 500 square foot area. Note if they are primarily conifers or deciduous, or if a mix is present.

2. Evaluate the understory. Determine if it is composed primarily of shrubs, or a mixture of shrubs and non-woody **herbaceous** plants.

3. Determine if grasses are present, and to what extent.

MAPPING TECHNIQUES

A "Quick and Dirty" Approach

As you walk through your 500 x 500 square foot area, look for differences in the plant communities present. Indicate on your map the approximate locations of groupings of plants, such as grassy areas, forested areas dominated by conifers, forested areas dominated by deciduous trees, mixed forest areas, shrub areas, etc. You may find that some forested areas are relatively mature while others are dominated by shrubs and immature trees. Sketch these observations on your map.

A More Detailed Approach

While still simple, this next approach enables you to map observations with more accuracy by actually taking distance measurements. Use the transects that you set up when you created your stream reach base map. Walk along each transect, stopping at regular intervals to observe, identify, and record the vegetation. If you make observations every 50 feet, you will get a good picture; for greater detail, make observations at closer intervals.

Using a compass in combination with a hip chain, tape measure, or pacing techniques, you can measure the distance and direction a particular plant grouping lies from a particular point along a transect. Then you can easily transpose the information onto your stream reach map. If you would like more detail, you can record the location of individual plants of interest, such as mature trees, dominant patches of shrubs, or special uncommon plants.

RIPARIAN ZONE WIDTH

In addition to describing the types and species of vegetation present, estimate the width of the riparian zone. The riparian zone is the area of vegetation affected by the flooding and high water table that occurs in the area next to the stream. In some cases, the riparian zone boundary is very clear; you should be able to distinguish a zone of vegetation along the stream that is different from the vegetation that occurs in the higher, drier ground further from the stream. Sometimes, however, a distinct riparian corridor is not evident.

The width of the riparian zone depends on a variety of factors, including the size of the stream, its flooding patterns, its interaction with groundwater and subsurface drainage, and the slope of its streambanks, floodplain, and valley.

1. Record your estimate of the riparian zone width on Data Sheet 2.

2. Delineate a rough boundary of the riparian corridor by measuring the distance from the stream to where the plant species seem to change to upland types. Record the boundary on your map. If you measure this distance at each transect, it will be easier to draw it on the map.

figure 4.15

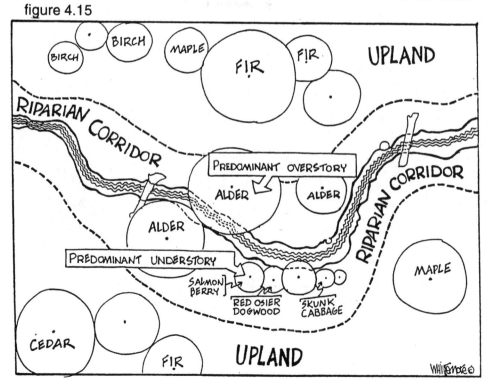

The Streamkeeper's Field Guide

FIELD PROCEDURE:
OVERHEAD CANOPY

While walking your stream channel, estimate the amount of the vegetation canopy overhanging or shading the stream. Do not consider vegetation that is less than 1/2 meter from the surface of the water. This way, you can distinguish vegetation that provides **overhead shade canopy** from vegetation that provides **overhanging protective cover**. Overhanging vegetation providing protective cover is generally within 2-3 feet of the water surface (see Field Procedure: Instream Habitat, on page 89.)

You can estimate overhead shade canopy from the bank, or be more accurate by standing (carefully) in the stream and looking up to estimate the percent of sky the canopy blocks from your view. If possible, make one estimate for the entire length of the stream reach, record this in the space provided on Data Sheet 2.

If the canopy cover is quite varied within the reach, make a separate estimate for each stretch between transect lines. In this case, record your estimates by making notations on your reach map. (The types of vegetation present should already be shown on your map and inclued with your riparian vegetation survey information.)

Illustration by Sandra Noel from *Adopting A Stream: A Northwest Handbook*, published by the Adopt-A-Stream Foundation, 1988

FIELD PROCEDURE: GRADIENT

Determine if the gradient of your stream reach is flat, moderate, or steep, and record your observation on Data Sheet 2. Flat gradient streams tend to have slow currents and slack water surfaces not broken by boulders or cobbles; the substrate is usually composed of fine sediments. In moderate gradient streams, the current is swifter. Riffles, bubbles, and/or white water are present because the water surface is broken by cobbles. In steep gradient streams, cascades, plunge pools, and waterfalls are present where the water flows over and around boulders and bedrock.

PERCENT GRADIENT

You can determine the gradient of your stream quantitatively by calculating the numerical and percent gradient. The numerical gradient of a stream between two points is calculated by dividing the elevation change between the two points by the horizontal distance between the points. The percent gradient is the resulting value multiplied by 100. Three methods for measuring the elevation change and distance between two points follow.

The first method is an estimate of the gradient of your entire stream reach, obtained from a topographic map. The second two are field methods that will yield gradient values for portions of your stream reach. Method III is more accurate than Method II because it involves using a level and enables you to calculate the gradient over a longer distance.

For more detail, you can use method II or III to measure the gradient of the whole stream reach in small, linked parts and plot the result on graph paper. This way you can produce a visual tool you can use to compare one reach with another.

Gradient Method I: Topographic Map

1. If you haven't already, obtain a topographic map of your stream reach area, with as large a scale as possible (a detailed, close-up view). Locate the beginning and the end of your stream reach on the map, and note their respective elevations.

2. To determine the distance between those two points, use a string to trace the path from the beginning to the end of the stream reach. Measure the length of string used and calculate the distance this length represents by comparing it with the scale in the legend of the map.

3. Use the formula below to calculate percent gradient of your stream reach, and record your result in Data Sheet 2.

Gradient Method I: Topographic Map
% Gradient = ($\blacktriangle$E / D) x 100
$\blacktriangle$E = elevation change = elevation of the beginning of your stream reach (upstream) minus elevation at the end of your stream reach (downstream)
D = distance between these two points

Gradient Method II: Line of Sight

1. One person stands downstream at point A and looks **straight and level** at a pole or stadia rod* held by a second person standing upstream at point B.

*Later on in your stream monitoring, you will need a **stadia rod**, which is a measuring stick or pole designed especially for stream surveys. See page 97 for information on making a stadia rod.

2. The person at point A notes the exact spot on the pole or stadia rod that their level line of sight reaches. The person holding the pole at point B then marks this spot and measures the height of the spot above the ground.

3. Use a tape measure or line held taut above the ground to measure the distance between points A and B. Make sure you do not measure the distance at ground level, because this will be greater than the true straight line distance between the two points, unless the gradient is very flat.

4. Use the formula on the next page to calculate the percent gradient between points A and B. Repeat this procedure for two other sections of your reach and then find an average value for the gradient of your stream reach. Record your result on Data Sheet 2.

The Streamkeeper's Field Guide

**GRADIENT METHOD II
LINE OF SIGHT**

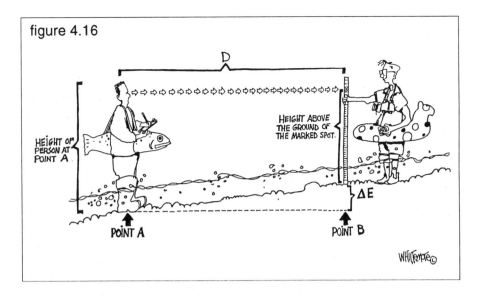

figure 4.16

Gradient Method II: Line of Sight

% Gradient = (▲E / D) x 100

▲E = elevation change = height of the person at point A minus the height above the ground of the marked spot at point B

D = distance between points A and B

Gradient Method III: Level Line

% Gradient = (▲E / D) x 100

▲E = elevation change = the height of the line at point A minus the height above the ground at point B.

D = distance between points A and B

Gradient Method III: Level Line

1. Tie a line to a fixed object at the upstream point of the section for which you'd like to calculate the gradient. If there is no appropriate object in the stream, have a second person stand in the stream holding a pole or stadia rod. Attach a line level to your line, move downstream to your desired endpoint, and pull the line taut.

2. Keeping the line level, measure the height of the line off the ground. Also measure the distance between the two points, which will be the length of the line between where you tied it and where you are holding it.

3. Use the formula (left) to determine the percent gradient of that section of your stream reach. Repeat this procedure for two other sections of your reach and then find an average value for the gradient of your stream reach. Record your result on Data Sheet 2.

**GRADIENT METHOD III
LEVEL LINE**

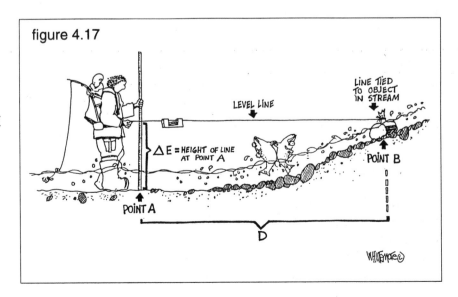

figure 4.17

Field Procedure: Sinuosity, Cross Section, and Stream Banks

SINUOSITY

Your mapping exercise will have given you an idea of the sinuosity of your stream, especially if you used 50' or smaller intervals between transects or if you used the mapping method II on page 71. This is the method which involves walking the stream channel from bend to bend with compass in hand. If you did not use this method and feel you would like to put greater detail on the meandering pattern of your stream on your map, you can go back and do parts 1 and 2 of mapping method II.

Use Data Sheet 2 to record if your stream is straight, meandering, or braided. You can calculate sinuosity by measuring the length of your stream and dividing that measurement by the length of the valley your reach is in.

CROSS SECTION

As you walk the length of your reach, evaluate the general cross sectional shape of your stream's **valley** as well as your stream's **channel**. Record your observations on Data Sheet 2 by checking the appropriate shape. You may want to do a more detailed analysis at specific study sites by plotting out the cross section profile of your stream's channel. The procedure for this can be found in Chapter 5, on pages 97-98.

STREAM BANKS

1. As you walk your stream channel, determine if vegetation on the stream banks is abundant, moderate, sparce, or non-existent. If you can, estimate the percent of the banks that are covered by vegetation. Record your observations.

2. Evaluate the stability of the banks for the length of your reach. Note on your map specific areas where banks are eroding or collapsed.

3. Note the extent and type of rockery and /or concrete placed along your stream reach. Evaluate the effectiveness of the protection by noting if it is associated with any erosion or collapsing of banks.

4. Observe the steepness of the banks along your reach. Try to estimate the percent of the reach that has banks with an incline of less than 45 degrees, between 45 and 90 degrees, and greater than 90 degrees (undercut). Record information.

FIELD PROCEDURE:
INSTREAM HABITAT

1. As you walk along your stream reach, count the number or pools, riffles, runs, and glides, and estimate the percent of the channel's length that is composed of each habitat type. Record your findings. Also try to document the location of the different habitat types on your stream reach map. Distinguishing between these four different habitat types can be a challenge.....nature tends not to make noticeable boundaries, especially in a stream. The most important thing is to be consistent. Look for locations where the water flow or turbulence changes. Usually this happens between habitat types. The size of the stream bottom substrate also tends to change, with larger materials found in runs and riffles and smaller materials found in glides and pools.

2. If you would like to record habitat types in more detail, measure the length of each pool, riffle, run and glide using a tape measure, hip chain, or pacing techniques. Start at one end of your reach and measure each habitat type, one by one. Then sum your values for each habitat type to find the total length of each. The stakes or tape you used to mark where transect lines cross the stream will help you with your estimations (they provide excellent reference points that you can locate on your map).

3. Calculate the **pool to riffle ratio** for your stream reach by dividing the number, percent, or length of pools by the number, percent, or length of riffles. Record your results. A ratio of 1:1 is desirable. However, if the total number or percent of pools and riffles are both low, you could still get a ratio of 1:1 and not have very good fish habitat. This would mean the reach was dominated by runs and glides, and thus not very diverse.

4. Note if large woody debris (logs, root wads, large branches) is abundant, moderate, sparse, or non-existent. Do the same for smaller organic debris, such as leaf litter and smaller branches. Sketch on your map the approximate location of these features.

5. Look for other features contributing to habitat diversity and protective cover, such as aquatic vegetation, overhanging vegetation, overhanging banks, and boulders. Overhanging vegetation that provides protective cover is typically less than 2-3 eet above the surface of the water. Note whether these features are abundant, moderate, sparse, or non-existent.

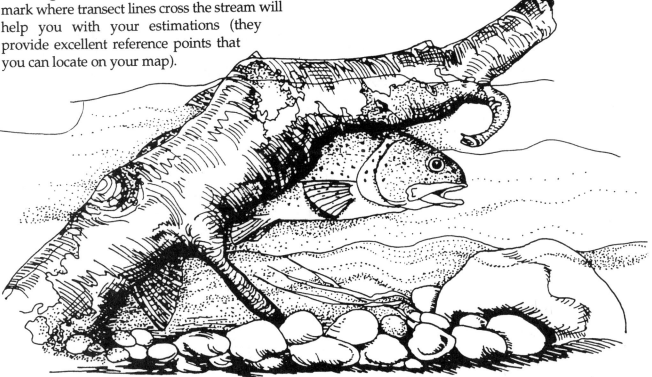

from *Adopting A Stream: A Northwest Handbook*; Illustration by Sandra Noel

FIELD PROCEDURE: HUMAN ALTERATIONS TO THE STREAM CHANNEL

1. As you walk the stream channel, check for any signs of human alterations. Look for signs of dredging, such as piles of sediments dumped on the banks. If the stream has been channelized, it may seem unnaturally straight and the banks may appear uniform in shape. If it has been diked, the top of the banks will be higher than the upland area just beyond the banks. Look for dams, weirs, and other ways the flow may be diverted, such as into culverts, irrigation ditches, or detention ponds. Record all your observations on Data Sheet 2 and note their approximate locations on your stream reach map.

2. Keep an eye out for garbage, soap suds, oil sheens, and other signs of pollution or toxic substances, and try to trace their sources. If you see pipes entering the stream, try to trace them back to their source to determine if they carry stormwater or point source pollution. If you determine a pipe is a stormwater drain, note if it comes from a detention system or directly from a road or parking lot. Record what you discover on Data Sheet 2 and on your stream reach map.

3. If you suspect that a culvert is blocking fish passage or causing erosion or flooding problems, contact a local water resources or fisheries agent for assistance. Consult Table 6 on page 114 for information on velocity barriers to migrating fish.

Problems Requiring Immediate Action

If you find a serious problem while surveying, such as an oil spill or a leak from a container that you suspect to be a possible toxic substance, report it to a local or state agency. Most states, counties and large cities have water quality "hotlines" which answer your questions, direct you to appropriate people, and relay your information onto field agents for investigation.

Agencies receive many phone calls. An agency field representative may be the sole person assigned to investigate a large area of the state, and actions to investigate or alleviate water quality problems may not always be as quickly or easily resolved. But, bear in mind that most water quality specialists have the same interest in healthy streams that you do and are concerned about your problem.

It is always best to have your problem investigated at the local level if possible. Start with the city's public works or surface water division. They will probably know more about a local problem than a state/provincial agency representative. What's more, they should be able to respond more quickly.

FIELD PROCEDURE: LAND USE

Use Data Sheet 2 to record the types of human land uses you see in your stream reach. To record the location of objects on your map, you can use the techniques outlined in the field procedure for surveying vegetation. The "quick and dirty" approach involves simply estimating the locations of objects you see during your survey. The more detailed approach involves measuring the distance from objects to points along your transects, using a compass in combination with a hip chain, tape measure, or pacing techniques. For more information on these techniques, see the field procedure for surveying vegetation on page 83.

NOW WHAT?

As you can see, conducting a stream reach survey is an involved process that can provide you with an abundance of information. Some Streamkeepers may decide to pick and choose certain procedures that best meet their goals. Once you start surveying on a seasonal basis, the process will become easier and require less time. With subsequent mapping and surveying, you will be able to detect changes that occur in your stream reach over time.

If you find that the stream reach has not changed much from the last time you surveyed, you can simply make notations on the last copy of Data Sheet 2 that you completed. If changes are more significant between seasons, use a separate copy of Data Sheet 2 each time you conduct your survey.

Each time you map your stream reach, you can create a new map or, instead, treat your first map as a base map and make overlays from each consecutive survey. To do this, sketch the new locations of your observations on a transparency that can be laid over the base map. Overlays are useful ways to see more visually how things have changed over time.

How do you use the information you gathered in your stream reach survey to evaluate the health of your stream? First look at the status of fish and wildlife. Compare your findings with historical data to determine if the populations seem to be on the decline. If so, it's likely you'll find clues that explain the decline in your other survey data. Has the riparian vegetation been altered? Is there no longer a tree canopy to keep the stream water cool? Are stream banks eroding? Has instream habitat been scoured out or silted in by stormwater runoff? Have people altered the stream channel?

Begin to draw connections between the land uses you observed during your survey and the observations you made about the stream itself. Reflect back on your earlier watershed detective work and recall what the land use zoning is within your stream reach and the surrounding watershed. You may discover some land uses that do not conform to the approved zoning. Highlight these areas on your stream reach map for later discussions with the local government in charge of land use.

But before you jump too far ahead, there's more monitoring to do. Chapters 5 - 7 will help you take a closer look at the physical, biological, and chemical aspects of your stream. You will learn how to put all your data to use in Chapters 8 - 10.

Name_____ Group_____ Date_____ Time_____

Stream Name_____ Reach Name/#_____

Section_____Township_____Range_____ Reach length_____

Reach Begins_____Ends_____ (in latitude/longitude or river miles)

Reach Landmarks_____

Driving/Hiking Directions_____

Weather Conditions ❑ Clear ❑ Cloudy ❑ Rain ❑ Other _____

Air temperature _____°___(C or F) Recent weather trends_____

Fish:

Type/Species (if known)	# Adults	# Juveniles	# Dead	# Redds or nests	Description & comments

Wildlife:

Birds			Herps (reptiles & amphibians)			Mammals	
type, species or track/sign	# or comments		type, species or track/sign	# or comments		type, species or track/sign	# or comments

Vegetation:

Type	Abundant	Moderate	Sparse	% of 500 ft² area covered	Species present
Conifer trees					
Deciduous trees					
Shrubs					
Herbaceous					
Grasses					

	0-50'	50-100'	100+
Width of Riparian Zone: Looking downstream: Left Bank	❑	❑	❑
Right Bank	❑	❑	❑

Overhead Canopy: (at least 3' above water) 0-25% ❑   25-50% ❑   50-75% ❑   75-100% ❑

Channel Characteristics:

Gradient Low ❑ Moderate ❑ Steep ❑ _____%

Sinuosity Straight ❑ Meandering ❑ Braided ❑ Ranking_____

Cross section shape:

Valley ❑ ❑ ❑   Channel ❑ ❑ ❑ ❑ ❑

Stream Banks:

Vegetation cover: Abundant ❑ Moderate ❑ Sparse ❑ ____%

Bank stability: Erosion in some areas ❑   Erosion in many areas ❑   Intact ❑
Collapsed in some areas ❑   Collapsed in many areas ❑

Artificial Protection: none ❑   < 25% ❑   25-50% ❑   > 50% ❑

Describe and evaluate_____

Bank steepness: (What percent of the total length is represented by each?)

<45°___%   >45°___%   90°___%   undercut > 90___%

Reach Habitat:

or length of pool_____ divided by # or length of riffle_____ = pool : riffle ratio _____

Large Woody debris:	Abundant ☐	Moderate ☐	Sparse ☐	None ☐
Small Organic debris:	Abundant ☐	Moderate ☐	Sparse ☐	None ☐
Overhanging debris:	Abundant ☐	Moderate ☐	Sparse ☐	None ☐
Overhanging banks:	Abundant ☐	Moderate ☐	Sparse ☐	None ☐
Overhanging vegetation:	Abundant ☐	Moderate ☐	Sparse ☐	None ☐
(< 3 ft. above water)				
Aquatic Vegetation:	Abundant ☐	Moderate ☐	Sparse ☐	None ☐
Boulders:	Abundant ☐	Moderate ☐	Sparse ☐	None ☐

Human Alterations:

Dredging ☐	Garbage/litter ☐	Culverts ☐
Channelization ☐	Toxic substances ☐	Pipes ☐
Diversions ☐	Sewage ☐	Detention ponds ☐
Dams ☐	Bridges ☐	Storm drains ☐
Weirs ☐	Roads ☐	Other_____ ☐
Dikes ☐	Other_____ ☐	Other_____ ☐

Land Uses:

(Enter 1 if present, 2 if you think the land use is impacting the stream.)

residential _____	forestry _____	grazing _____
commercial _____	mining _____	crops _____
industrial _____	recreation _____	irrigation _____

Comments On Stream Reach:

## CHAPTER FIVE

Monitoring Your Stream's Physical Characteristics

By this time you have inventoried your watershed and mapped and surveyed your stream reach. The next three chapters cover the stream monitoring parameters that you will be assessing at specific **study sites** within your stream reach. In this chapter, you will learn to analyze some physical parameters about your stream in greater detail: the stream channel cross section profile, the flow dynamics of water in the channel, and the nature of the stream bottom.

The criteria you use to select sites will depend on the particular parameter you are measuring. No matter what the parameter, however, it is important that you mark the site in the field and document the site location on your data sheets and in your Quality Assurance Plan. This will help make repeat measurements in the same places for long term comparison studies. Refer to the section "Documenting Locations," on page 74 in Chapter 4.

Also, remember to consult "Miss Mayfly's Guide to Watershed Etiquette, Safety, and Liability" before you head out into the field to survey your study sites.

CROSS SECTION PROFILE

As part of your stream reach survey, you evaluated the overall cross section shape of your stream. If you want more detailed information about cross section shape, you can plot the cross section profile of your stream channel. This involves taking depth measurements at specific locations along the channel and plotting them on graph paper to create a visual diagram of the cross section. Making such a plot gives you something more concrete to track through the seasons than the visual evaluation of cross sectional shape that was part of the stream reach survey.

The procedures in this section show you how to plot the profile of the stream channel at its **bankfull flow stage.** Bankfull flow refers to high flows that fill the entire channel up to the top of the banks, to the level of the floodplain. When a stream or river reaches its bankfull flow stage, it is at the point where it begins to overflow onto its floodplain. This stage is obviously important for understanding the dynamics of flow and flooding in stream and river systems.

figure 5.1 Cross Section Profile

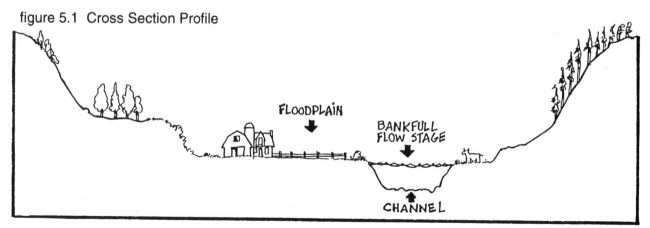

FLOODPLAIN

BANKFULL FLOW STAGE

CHANNEL

WHITTEMORE©

FIELD PROCEDURE: CROSS SECTION PROFILE

EQUIPMENT NEEDED

pencils
clipboard
Data Sheet 3, "Cross Section Survey"
site markers (stakes, forestry tape, etc.)
improvised stadia rod (see window)
tape measure
carpenter's level
string
waterproof boots/waders

WHAT?

In this procedure, you will survey the stream channel cross section so you can plot its profile on graph paper. You will be able to plot both the bankfull flow profile and the profile of the **wetted stream**, where water is present.

WHERE?

We recommend you do a profile plot at locations where each of the four different habitat types are present, ie., do a profile plot at a pool, at a riffle, at a run,* and at a glide. You may also want to do profile plots at certain "trouble spots" that you identified in your stream reach survey. A trouble spot might be a place where you detect erosion, unstable banks, or land uses that look like they may affect the shape of the channel and banks. When you can, use the points where your transects cross the stream so it's easier to return to the sites for future surveys.

* You can save yourself a little time if you use the same run where you plan to measure flow, but only if you'll be able to measure flow on the same day that you conduct your profile survey. (See the Flow section for more information).

Mark your stream channel profile sites in the field and document their locations on your data sheet and in your Quality Assurance Plan to ensure that you compare the same places over time. Describe the sites, including what type of habitat they represent.

figure 5.2
Making an Improvised Stadia Rod

Use a wooden dowel or metal rod (such as a rebar rod or metal fence post). Paint it with alternating six-inch-long sections of bright red and white. After it dries, paint black lines over the red and white, marking 0.1 foot intervals. This will give you a handy measuring device that is a hybrid between Metric and English (American) systems. You can use the rod to read out length measurements in feet and tenths of feet at the same time.

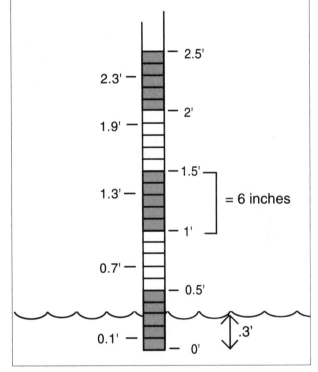

WHEN?

Try to plot the cross section profile each time you survey a stream reach, which is recommended four times a year, once each season.

HOW?

It works best if you have at least three Streamkeepers to carry out this procedure (see figure 5.3).

1. Stretch a tape measure across the stream at the height of bankfull flow and perpendicular to the stream flow. The height of bankfull flow is the top of the stream banks, the level of the floodplain. Note that the height of bankfull flow

is not the same as the wetted edge of the stream unless the stream is at its maximum flow before overflowing onto its flood plain. (In most cases, it will not be safe to survey your cross section when your stream is at bankfull flow.)

2. Use string to tie the tape measure to a tree or stake at the top of one bank. The tape measure should lie level with the top of the bank. (If the two banks are not of equal height, tie the tape to the lower bank.) The 0-point on the tape measure should be even with the edge of the top of the bank, just as it starts to slope downward toward the stream. From the opposite bank, hold the tape measure level and taut. Attaching a carpenter's level to the tape measure will result in greater precision.

3. Station another person on the bank, clipboard and Data Sheet 3 in hand, to record measurements.

4. A third person, stadia rod in hand, moves along the tape measure at one-foot intervals and measures the vertical distance from the ground to the tape measure at each interval. This distance will be the **bankfull flow depth** of the stream channel at each point.
Note: if the channel is more than 20 feet wide, measure vertical distances at 2-foot intervals.

5. The stadia rod person continues measuring until reaching the first **wetted edge** (the point where the water surface hits the bank). Whether or not the wetted edge falls on an interval, measure the horizontal distance from the starting point to the wetted edge. The clipboard person records this information on Data Sheet 3.

6. As they proceed toward the far bank, the stadia rod person should now measure the **wetted stream depth** in addition to the bankfull flow depth. The wetted stream depth is the distance from the bottom of the stream to the surface of the water, while the bankfull flow depth is the distance from the bottom of the stream to the tape measure.

7. The stadia rod person works toward the far bank, measuring the bankfull flow depth and the wetted stream depth at each interval.

8. When they reach the second wetted edge, the stadia rod person measures the horizontal distance from wetted edge to wetted edge.

9. The final measurement recorded on Data Sheet 3 is the horizontal distance of the ending point from the starting point. The ending point is the point where the tape measure hits the far bank. The horizontal distance is the **width of the bankfull flow** stage.

10. Plot out the vertical depth measurements and the horizontal distance measurements on graph paper. On the plot, indicate the location of the bankfull flow, water level, wetted edges, and bank edges. The result will be an accurate cross sectional plot of your stream that will give you a good sense of the shape of the stream channel. Attach this plot to Data Sheet 3.

NOW WHAT?

Observe how the profile of your stream channel changes over time and in response to major storms or human-related events.

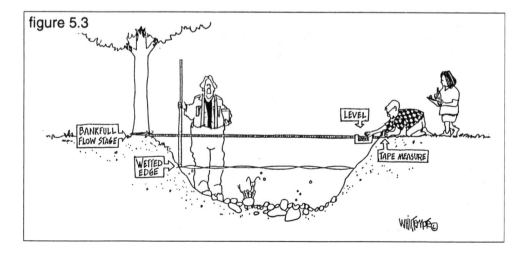

figure 5.3

BANKFULL FLOW STAGE

WETTED EDGE

LEVEL

TAPE MEASURE

DATA SHEET 3: CROSS SECTION SURVEY

Name _____ Group _____ Date _____ Time _____

Stream Name _____ Reach Name/# _____

Weather Conditions: ☐ Clear ☐ Cloudy ☐ Rain ☐ Other _____

Air Temperature: _____°_____ Recent weather trends: _____

Site Name/# _____ Site Description: habitat type, landmarks, etc. _____

Site Location: Latitude ___°___'___"N Longitude ___°___'___"W River mile: _____

Driving/Hiking Directions _____

Depths at 1 ft Intervals:

distance (1 ft. intervals)	1	2	3	4	5	6	7	8	9	10	11	12	13	14	15	16	17	18	19	20
bankfull depth (ft)																				
wetted stream depth (ft)																				

Horizontal Distance:

Point	Left bank start point	Left bank wetted edge	Right bank wetted edge	Right bank start point
Distance	0 ft			

THE STREAM BOTTOM

The stream bottom is extremely important to the organisms that live there. Some crucial activities that take place here are:

- Growth of bacteria, fungi and benthic algae that feed macroinvertebrates;
- Growth and development of bottom-dwelling macroinvertebrates that feed fish and invertebrates;
- Fish reproduction and egg development;
- Rearing of juvenile fish, frogs, and salamanders.

The composition of the stream bottom varies over the length of the stream, from headwaters to mouth, from riffles to pools, from high to low flow areas, and from season to season. The bottom composition dictates which organisms live in the stream and indicates the natural forces and human impacts that have been at work.

The stream bottom is composed of particles and organic matter collectively called **substrate**. Substrate size ranges from the finest sediments such as silt, clay, and mud, to increasingly larger particles: sand, gravel, cobble, boulders, and solid bedrock. Burrowing organisms like worms and midges prefer the mud and sand bottoms of pool habitats, but salmonids and the majority of stream macroinvertebrates require a well-oxygenated gravel and cobble bottom.

Excessive amounts of fine sediments can impact aquatic life throughout the stream food chain. When sediment is suspended in the water column, the increased **turbidity** (low clarity) may prevent sunlight from reaching photosynthesizing algae and other aquatic plants. Turbidity can prevent fish from finding their prey and tends to clog the gills and filter feeding mechanisms of fish and various invertebrates.

Excessive sediment also increases the erosive and scouring force of water in a stream during high flows. During low flows, excessive sediment may be deposited on the stream bottom, filling the spaces between cobble and gravel particles. When this happens, buried fish eggs and stream organisms can be "smothered" because their oxygen supply is cut off. Important pool habitat may also be filled in.

Cobble embeddedness is the extent to which cobbles are surrounded or covered by fine

sediment. In general, when cobble embeddedness is 30-40%, salmonid spawning habitat is lost and macroinvertebrate populations are threatened.

In addition to causing cobbles to become embedded, fine sediments can also have a cementing effect, causing cobbles to be tightly compacted. Compaction of substrate can also be caused by high flows that scour the stream bottom, a frequent problem in our urban streams where storm water runoff is high. **Consolidation** is a measure of the compaction of the substrate. A highly consolidated streambed provides little or no fish spawning or macroinvertebrate habitat.

The composition of a stream bottom is always changing because substrate components move from the force of stream flow. This movement is referred to as **bed load transport**. The size of the substrate and amount of fine sediments present affects the amount of bed load transport. Larger substrate, like boulders and cobbles, are harder to move than sand and silt.

All streams require a degree of bed load transport to maintain their pools, riffles, and meanders. Some substrate movement is beneficial because it allows fine sediments to be flushed out of the spaces between larger particles and ultimately downstream. However, if there is too much substrate movement, the channel may be too unstable to support healthy fish and invertebrate populations. In general, an **optimum stability level** requires that less than 50% of the stream bottom material is mobile during bankfull flow conditions. (Recall that bankfull flow refers to high flows that fill the entire channel up to the top of the banks).

The Streamkeeper's Field Guide

FIELD PROCEDURE:
STREAM BOTTOM SURVEY

EQUIPMENT NEEDED

pencils
clipboard
Data Sheet 4, "Stream Bottom Survey"
stakes or forestry tape
meter, yard stick, or ruler
waterproof boots/waders

WHAT?

The stream bottom survey involves evaluating the sizes of the materials that make up the substrate, the amount of cobble embeddedness, and the degree of consolidation.

WHERE?

As with the cross sectional profile, we recommend you survey the stream bottom at locations where each of the four different habitat types are present, ie., survey the stream bottom of a riffle, a run, a glide, and if possible, a pool.

You can save yourself some time if you use the same riffle where you plan to collect macroinvertebrates, because the macroinvertebrate procedure requires that you analyze the stream bottom at your collection site. However, you must sample the critters on the same day that you survey the stream bottom. Also, because collecting the critters disturbs the stream bottom,

evaluate substrate size before you collect the critters, and estimate embeddedness and consolidation while you are in the process of collecting them. (See Chapter 7 for more information).

You may also want to survey the stream bottom at important spawning areas and at certain "trouble spots" that you identified in your stream reach survey. A trouble spot might be a place where you detect erosion, unstable banks, or land uses that look like they may affect the condition of the stream bottom. Try to use the points where your transects cross the stream so it's easier to return to the sites for future surveys.

Mark your stream bottom survey sites in the field and document their locations on your data sheet and in your Quality Assurance Plan. Describe the sites, including what type of habitat they represent. Make a note if this site is the same location where you collect macroinvertebrates

WHEN?

We recommend you survey the stream bottom each time you conduct a stream reach survey - four times a year, once per season. You will also need to do a stream bottom survey when you collect macroinvertebrates.

HOW?

Substrate Size

For each site that you survey, estimate the percent of the stream bottom that each of the differently sized materials represents. For example, if the site you are evaluating seems to be

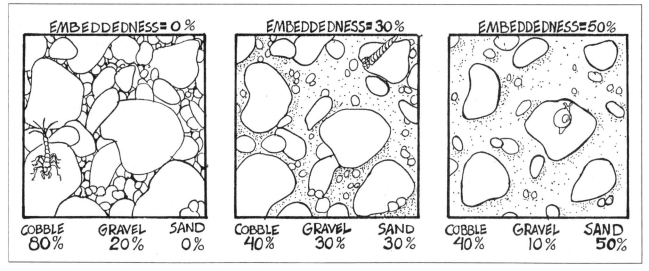

EMBEDDEDNESS= 0 %			EMBEDDEDNESS= 30%			EMBEDDEDNESS=50%		
COBBLE	GRAVEL	SAND	COBBLE	GRAVEL	SAND	COBBLE	GRAVEL	SAND
80%	20%	0%	40%	30%	30%	40%	10%	50%

covered with an equal mixture of cobbles and gravel, with a small amount of sand and silt, your substrate evauation may result in 10% silt, 10% sand, 40% gravel, 40% cobble, 0% boulders, 0% bedrock. Substrate size categories are listed in the window below, and also on Data Sheet 4. Use Data Sheet 4 to record your observations.

Substrate Size Categories

Substrate size categories, from smallest to largest, are as follows:

silt/clay/mud	
sand	< 0.1 inch
gravel	0.1 - 2 inches
cobble	2 - 10 inches
boulder	> 10 inches
solid bedrock	

Cobble Embeddedness

To estimate the percent of embeddedness within each site that contains cobble:

1. Select a cobble and remove it from the streambed, retaining its spatial orientation as you pick it up.

2. Estimate the percent of the cobble's height that is embedded by finer sediments. Usually you can see a line on the cobble marking the point where it emerges from the sediment to be exposed to the flowing water. If this line is about half-way down the height of the cobble, the embeddedness is 50%. If the mark is one-third of the way down, embeddedness is about 30%.

3. Repeat this analysis for several cobbles. If there are few cobbles visible, check to see if you can find any underneath the fine sediments.

figure 5.4

Any cobbles that are completely buried are 100% embedded.

4. Based on your observations, use Data Sheet 4 to record an average estimate of percent embeddedness for each plot.

Consolidation

To characterize how consolidated or hard-packed the substrate is at each site, simply try to move it with your boot heel while standing in the stream. Determine whether it is loose (easily moved), moderately difficult to move, or tightly consolidated (difficult to move by kicking). Record your results on Data Sheet 4.

Evaluating Consolidation

When you are evaluating consolidation with your boot heel, think about a female salmon trying to move gravel to make a nest. If the stream bottom is loosely consolidated, she will have an easy time. If the stream bottom is tightly cemented, she will not be able to dig a proper redd by moving her tail.

NOW WHAT?

With quarterly stream bottom surveys, you'll have a good record of how the substrate of your stream changes over time. You can also record how the substrate changes in response to major storm events, or in response to different land uses, such as a construction event, by doing "before and after" surveys at appropriate times.

Remember that you would not expect pool or glide habitats to be free from sediment, but runs and riffles should be. Keep an eye out for important spawning areas. If cobble embeddedness reaches 30-40% in those areas, salmonid survival will be threatened.

COBBLE EMBEDDEDNESS

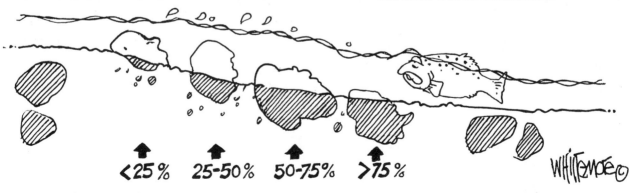

<25%   25-50%   50-75%   >75%

DATA SHEET 4 STREAM BOTTOM SURVEY

Name________________ Group________________ Date__________ Time__________

Stream Name____________________ Reach Name/#____________________

Weather Conditions: ❐ Clear ❐ Cloudy ❐ Rain ❐ Other ____________

Air Temperature: ______°___ Recent trends: ____________________

Site Name/#____________

Site Description (habitat type, landmarks etc.)________

____

Site location (lat & long, transect #, and/or river mile)

____

Substrate Size
(percent stream bottom that each size represents)

silt, clay, mud	
sand < 0.1"	
gravel 0.1-2"	
cobble 2-10 "	
boulder > 10"	
solid bedrock	

Embeddedness

(an estimate of the average percent that cobbles are embedded)

______ %

Consolidation

❐ Loose

❐ Moderately difficult to move

❐ Tightly cemented

Site Name/#____________

Site Description (habitat type, landmarks etc).________

____

Site location (lat & long, transect #, and/or river mile)

____

Substrate Size
(percent stream bottom that each size represents)

silt, clay, mud	
sand < 0.1"	
gravel 0.1-2"	
cobble 2-10 "	
boulder > 10"	
solid bedrock	

Embeddedness

(an estimate of the average percent that cobbles are embedded)

______ %

Consolidation

❐ Loose

❐ Moderately difficult to move

❐ Tightly cemented

Site Name/#_____

Substrate size
(percent stream bottom that each size represents)

silt, clay, mud	
sand < 0.1"	
gravel 0.1-2"	
cobble 2-10 "	
boulder > 10"	
solid bedrock	

Site Description (habitat type, landmarks etc.)
_____

Site location (lat & long, transect #, and/or river mile)
_____

Embeddedness
(an estimate of the average percent that cobbles are embedded)

%

Consolidation

❑ Loose

❑ Moderately difficult to move

❑ Tightly cemented

Site Name/#_____

Substrate Size

(percent stream bottom that each size represents)

silt, clay, mud	
sand < 0.1"	
gravel 0.1-2"	
cobble 2-10 "	
boulder > 10"	
solid bedrock	

Site Description (habitat type, landmarks etc.)
_____

Site location (lat & long, transect #, and/or river mile)
_____

Embeddedness
(an estimate of the average percent that cobbles are embedded)

%

Consolidation

❑ Loose

❑ Moderately difficult to move

❑ Tightly cemented

Comments

STREAM FLOW

Flow or **discharge** is the volume of water moving past a point in a unit of time. Two components make up flow: the **volume** or amount of water in the stream, and the **velocity** or speed of the water moving past a given point. Flow affects everything from the concentration of various substances in the water to the distribution of habitats and organisms throughout the stream.

How Flow Affects a Stream

The concentrations of **pollutants** and **natural substances** are influenced by stream flow. In larger volumes of faster moving water, a pollutant will be more diluted and flushed out more quickly than an equal amount of pollutant in a smaller volume of slower moving water.

Flow also affects the amount of oxygen dissolved in the water, as well as the temperature of the water. Higher volumes of faster moving water, especially if "white water," helps churn atmospheric oxygen into the water. Smaller volumes of slower moving water can heat up more dramatically in the heat of the summer sun.

Stream flow interacts with the gradient and substrate of a stream to determine the nature of **physical features**, such as the types of habitats present, the shape of the channel, and the composition of the stream bottom. Flow can determine whether an area is a run, riffle, glide, or pool. As depicted in figure 5.5 below, riffles can become

runs and pools can become glides when covered by higher flows.

The amount of **sediment** and **debris** a stream can carry also depends on flow. A large volume of fast moving water carries more sediment and larger debris than a small volume of slow moving water.

High volume, sediment-bearing flows have greater erosional energy, while smaller and slower flows allow sediment to be deposited. The alternating erosional and depositional activities of flowing water help to determine stream channel shape and sinuosity.

Flow also dictates whether a stream will overflow its banks and when floodwaters will drop their load of sediment and debris on an adjacent flood plain.

Because of its effect on the chemical and physical nature of streams, flow helps determine what **plants** and **animals** live in a stream. For instance, salmonids require high concentrations of dissolved oxygen, low water temperatures, gravel substrates for spawning, and quiet pools for rearing. Carp and catfish, on the other hand, will survive quite nicely in water with muck and lower oxygen levels.

Most species of anadromous fish also need specific flow volumes and velocities for spawning migrations and downstream migrations of juveniles.

Aquatic insects have similar physical and chemical requirements. Unlike fish, however they can't migrate and are captives of local stream conditions. They have adapted to specific flow regimes and types of stream habitats.

figure 5.5

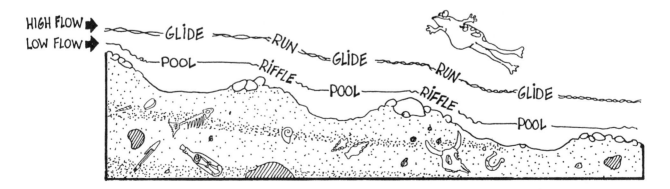

Suction disks and silk strands enable them to attach to rocks while ingenious feeding tools like nets and fans allow them to capture particles from flowing water. They also have streamlined and flattened body shapes which minimize their resistance to flowing water. If the specific flow regimes to which these organisms have adapted are altered, the organisms may suffer.

In areas where water is in short supply, governments may set minimum **instream flow requirements** for streams that are subject to water withdrawals. Out-of-stream uses such as domestic water use, hydropower generation, and irrigation, may leave little water for fish at crucial stages of their lives. Minimum instream flows are sometimes established in order to maintain valuable fish populations and to balance competing out-of-stream uses.

Factors Affecting Volume of Flow

The volume of stream flow is determined by many factors. **Precipitation** is generally the key factor, as it usually provides the primary source of water. After a rainstorm, stream flow follows a predictable pattern in which it rises sharply in response to the storm, and then falls, usually more gradually, in the hours or days following the storm.

This pattern is depicted in the storm hydrograph on the right. A hydrograph is simply a graphical way to illustrate flow by plotting the volume of water in a stream or river over a certain duration of time. A storm hydrograph does this for the time that elapses during and just following a storm.

Usually the highest flow resulting from a storm does not happen at the same time as the highest rainfall intensity of the storm. The hydrograph shows a **lag period** between the time the storm reaches its highest intensity and the time the stream reaches its peak flow - it takes the stream a little time to "catch up" with the rainfall.

The hydrograph also illustrates the level of **base flow**. Base flow is simply the volume of water in a stream before the effect of a storm kicks in. It is shown on the hydrograph as the shaded area. The amount of **runoff** that occurs after a storm is the total amount of flow resulting from a storm minus the base flow. Runoff is indicated on the graph by the area under the curve that is above the shaded base flow area.

While flow changes over a period of a few hours in reponse to rainstorms, it also varies seasonally. Seasonal variation in precipitation changes the timing and quantity of runoff to streams. Usually streams have predictable periods of maximum and minumum flows that coincide with wet and dry seasons.

Vegetation also affects the volume of streamflow. It absorbs water, releasing it to the atmosphere through evapotranspiration. It also increases the water storage capacity of soil, helping

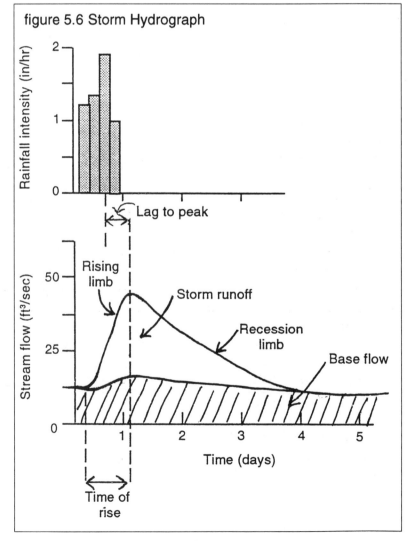

figure 5.6 Storm Hydrograph

to conserve moisture in dry times. Springs, lakes, adjacent wetlands, and tributaries also affect the volume of stream flow. While each source may contribute only a small portion of the total flow in a stream, their contributions can be crucial during dry seasons.

Factors Affecting Velocity of Flow

The velocity of flow in the stream channel is affected by a variety of factors. **Gradient** is key, as water moves faster in steeper areas. **Resistance** is another important factor. Velocity is slower in areas of high resistance. Resistance is determined by the nature of the bottom substrate, channel shape, woody debris such as logs and root wads, and stream habitats such as riffles and pools. A meandering stream with large objects in the stream and on the stream bottom resists flow more than a straight stream with a smooth bottom.

How Human Activities Affect Flow

Stream flow can be altered by human activities both in the surrounding watershed and directly in the stream. When naturally vegetated areas and wetlands are converted into bare soil and impervious surfaces, the volume of stream flow will increase during wet seasons. In addition, the lag period will decrease, as discussed in reference to the storm hydrograph on page 106, i.e, the peak flow after a rainstorm will happen more rapidly in watersheds where vegetation and wetlands are removed.

Thus, during wet periods, vegetation and wetland loss results in too much water running off too quickly. In dry times, however, the opposite problem occurs. Without natural vegetation or wetlands, much of the water storage capacity of a watershed is lost. In dry times, stream flow may be severely reduced or even non-existent.

Channelizing a stream and removing woody debris and other large objects can increase the velocity of flow. In the past, it was a common practice to straighten and "clean up" streams, in order to prevent flooding by facilitating the movement of water quickly downstream.

More recently we've realized that these practices may actually create worse problems than they solve. Besides the serious consequences they have for aquatic and riparian life, these practices often cause erosion and flooding problems in downstream locations. A straight, "clean" stream channel has much less ability to dissipate and absorb the force and volume of flow. Nature has created messy, meandering streams for a good reason.

Dams Are Not a Fish's Best Friend:

The Columbia River salmon runs have plummeted to 15-20% of their original populations. While many factors have contributed to this decline, the hydroelectric dams pose a significant problem both for upstream swimming adults and downstream swimming juveniles.

The Kootenai River sturgeon have not successfully spawned since the construction of the Libby Dam in Montana in 1972. These long-lived fish were listed as a federally endangered species in 1994.

Dams or diversions change the flow of water in stream channels directly by slowing it, detaining it, or re-routing it. Hydroelectric dams can cause flows to fluctuate greatly at times when power is needed, dramatically altering the physical and chemical conditions of rivers. Some aquatic organisms that have timed their life cycles to natural flow regimes are able to adapt to the new situations; but many are seriously affected by the highly altered systems created by dams or diversions that each size represents.

FIELD PROCEDURE: FLOW

EQUIPMENT NEEDED

pencils
clipboard
Data Sheet 5, "Flow"
tape measure (50 feet)
markers (string, stakes, or forestry tape)
visible float (orange, ping-pong ball, stick, pine cone, rubber ducky, or whatever works - see "Velocity Float" on page 109)
stopwatch
calculator
cross section profile plot or equipment for surveying cross section (see section earlier in this chapter)
waterproof boots/waders

WHAT?

Flow is the volume of water that passes a point in the stream in a unit of time:

flow = volume / time

If your stream flows through a culvert with a small amount of water such that all the water can be collected in a bucket or other container, you can actually measure flow directly by measuring the volume of water that flows in a set period of time. See the watershed window on page 113 for instructions on this "short cut" method.

In most cases, however, it is impossible to measure the volume of water in a stream directly. Thus, you should break flow down into measurable components. Since volume is a product of length and cross sectional area,

flow = (length x cross sectional area) / time

And since velocity is a measure of length (or distance) /time,

flow = velocity x cross sectional area

DEADLY TURBINE BLADES ON THE WAY DOWN... PHYSICAL BARRIERS ON THE WAY UP... KIDS, *EITHER WAY*...

YEAH, WE KNOW... DAMMED IF YOU DO, DAMMED IF YOU DON'T!

The Streamkeeper's Field Guide

To calculate flow, you will need to measure the velocity of the water and the cross sectional area of the stream's channel.

The procedure outlined in this book is a simple method requiring inexpensive equipment. Rather than measuring velocity with a $1000 flow meter, you will use a **float** (orange, ball, stick, rubber ducky, etc.).

While the float method serves more as an estimation of flow, it is a tried and true method. It may even work even better than a meter under very low flow conditions. With proper quality assurance your results can be accurate and useful. (See the "Now What" section on page 114 and page 200 in Chapter 8, "Credible Data," for more information on quality assurance and quality control).

In addition to measuring flow, we also recommend you keep track of storm events, especially those events that occur in the week preceding your flow measurement. At a minimum, you should keep track of the amount of precipitation that falls and the time that elapsed between the last storm and your flow measurement.

What is the Best Velocity Float?

We have found that oranges work well. Because they float partially submerged in the water, they are less subject to wind resistance. Because they are fairly water-proof, they do not absorb water and get heavier with each float trial. (You can also eat them when you are done.)

In slow moving or shallow water, a light object like a twig or pine cone may work best. Ping pong balls are easier to see than twigs and pine cones but they are more prone to wind resistance. Try out different techniques and determine which gives you the most precision. Then be sure you document what type of float you use on your Quality Assurance Plan.

WHERE?

You will need a section of your stream reach that is as straight and uniform in width as possible (channelized sections of streams work well for this exercise). The section you choose should be free of turbulence, eddies, slack backwater areas, and obstacles such as boulders and woody debris.

The section you choose should also be shallow enough for you to wade across safely. In general, the length of the section should be about three times the width; however, do not sacrifice the other conditions listed above if this is not possible. A ten foot or even shorter distance will work.

If your stream is culverted for a short distance within your reach, check to see if a float will flow freely through the culvert without interruption or turbulence. If so, we recommend you use the culvert for measuring flow. Because culverts have very uniform dimensions, you will encounter less variability and be able to take measurements quickly and easily.

Mark the locations of your flow sites in the field and document these locations on your data sheet and in your Quality Assurance Plan to ensure that you come back to the same place every time you measure flow.

WHEN?

Because flow is so variable, we recommend that you measure it at least once per month. In order to obtain a comprehensive picture of your stream's flow throughout the year, take a flow measurement on approximately the same day each month.

You may consider evaluating the effect of storm events on the flow of your stream. This is more labor intensive than monthly observations, because for each storm, you will have to take flow readings at regular time intervals from the beginning of the storm until some time after it ends. However, it may be important information to gather if you are concerned about increased runoff volumes resulting from development in your watershed.

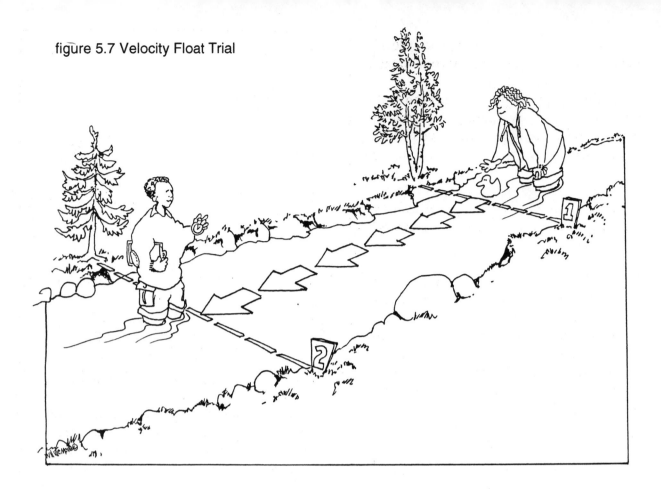

figure 5.7 Velocity Float Trial

HOW?

Velocity

You will need at least two people, and best three, for this exercise (see figure 5.7, above).

1. Measure the length (distance) of the stream section you have chosen for your flow measurement and record it on Data Sheet 5, "Flow." Mark the start and finish lines in some manner (a tape measure or string, held across the stream perpendicular to the direction of flow, works well).

2. Person #1 wades in the stream at the upstream starting line, float in hand. Person #2 wades in the stream at the downstream finish line. Person #3, if available, stands on the bank next to the finish line, stopwatch and clipboard in hand. If you have only 2 people, the one at the fish line (#2) holds the stopwatch.

3. Person #1 drops the float on the surface, upstream from the starting line. As the float passes the starting line, person #1 yells "go" and

person #2 starts the stopwatch. (It is more accurate to start the float this way than trying to drop it exactly at the starting line.)

4. When the float crosses the finish line, person #2 stops the stopwatch, catches the float, and records the time, or gets person #3 to help out.

5. Discard any trials in which the float gets caught in debris, rocks, or eddies.

6. On Data Sheet 5, record the time in seconds that it took for the float to travel the measured distance.

7. Because the velocity of the water varies across the width of the stream, you need to repeat this process several times, sending the float down different flow paths. Starting at one side of the stream, send the float down progressively farther from the bank, at one or two foot intervals until you reach the other side. This way you will be able to sample the full range of velocities that occur across the width of the stream. We

recommend ten float trials to get a velocity measurement representative of your stream.

8. Record all the time values on Data Sheet 5. Calculate the average float time by dividing the sum of the time values by the number of float trials.

9. To calculate the average velocity, divide the distance value by the average float time value. Your result is the **average surface velocity.**

10. Because the velocity of a stream varies from the surface to the bottom, adjust your result to reflect the overall average velocity of the stream. (See inset on page 112 for the formula and a detailed explanation). This adjusted value is called the **corrected average velocity.** Record your result on the data sheet.

ETHEL BREAKS THE VELOCITY BARRIER BUT FINDS SHE IS UNABLE TO CHANNEL HER BURST SPEED.

WHITTEMORE ©

For more information on burst speed, see Table 6 on page 114

Cross Sectional Area

This section outlines two methods for measuring and calculating cross sectional area. The first method is more accurate for streams wider than three feet, but requires that you take depth measurments in one (or two) foot intervals across the width of your stream. If it is too difficult to do this on your stream, or if your stream is only a few feet wide, method II should work better for you.

If you have surveyed the cross section profile at the same site and on the same day as your flow measurement, you can save yourself a little

time by using data from your profile survey. The sum of the **wetted stream depth measurements** you took at one foot intervals equals the cross sectional area of the wetted stream at the point surveyed. Find this value, and then use Method I on page 112 to find the cross sectional area at two more points along the length of the channel you used for the velocity float trials. Use these three values to calculate the average cross sectional area of that section of the channel.

If you used a culvert to measure velocity, simply use method II to calculate the cross sectional area of the culvert. You need not find an average (assuming the culvert is of uniform size from beginning to end).

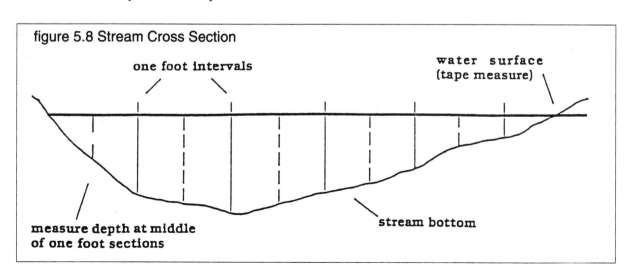

figure 5.8 Stream Cross Section

one foot intervals

water surface (tape measure)

measure depth at middle of one foot sections

stream bottom

Method I
for Streams More than Three Feet Wide

1. Stretch a tape measure across the stream at the starting point of your velocity float trials. The tape measure should be perpendicular to stream flow and just above the surface of the water. The first person holds the "0" point on the tape measure even with one wetted edge. A second person on the opposite bank holds the tape measure level, taut, and even with the other wetted edge. A third person stands on the bank with Data Sheet 5 and records information.

2. A fourth person, **stadia rod** in hand, starts six inches from wetted edge and moves along the tape measure at one foot intervals. The stadia rod is used to measure the depth of the water (the vertical distance from the water surface to the stream bottom) at the **mid-point** of each one foot interval. (Thus, the first and last measurements are made 6 inches from the wetted edges.) The in-stream person calls out these measurements to the person on the bank, who records them on Data Sheet 5. (See page 97 for information on stadia rods.)

Note: If your stream is wider than 20 feet, measure depths at two-foot intervals.

3. If measured at one-foot intervals, each depth is roughly equivalent to the area of a one-foot-wide section of the wetted stream. Thus the sum of these areas (or depths) is the total cross-sectional area of the wetted stream at the point measured. The mid-point depth is used because it is a better approximation of the area of the one-foot-wide section than the depth at the beginning or end of the section.

4. Sum the depths and record the result for Cross Section 1 on Data Sheet 5. If you measured depths at two foot intervals, multiply the sum of the depths by two to calculate cross sectional area.

About the Velocity Correction Factor

The velocity of a stream varies from surface to bottom. While average stream velocity can be found at a point slightly deeper than stream center; the resistance of the stream bottom decreases velocity near the stream bottom and surface velocity is about 16 percent faster than at the center.

Since you measure surface velocity with the float method, you need to multiply your result by a correction factor to get the average velocity. The correction factor should decrease your value, since the surface velocity is faster than the average. Thus the correction factor will be less than 1.

Thomas Dunne and Luna Leopold, internationally recognized hydrologists and authors of *Water in Environmental Planning*, use a correction factor of 0.8 when calculating flows with the float method. This enables them to convert surface velocity to average flow velocity. Follow their example for more accurate results.

5. Your result is the cross sectional area of the wetted stream at the starting line of your velocity float trials. In order to find the average cross sectional area of the whole length of the channel used for the float trials, you'll need to repeat steps 1-4 at two more points. Use the finish line and a point in the middle of the section. This is especially important if the area of the channel varies along the length used for the float trials.

6. Take the three cross sectional area values and average them to find average cross sectional area. Record the result on Data Sheet 5.

Method II
For Streams Three Feet Wide or Less

1. Stretch a tape measure across the stream at the starting point of the channel you used for the velocity float trials. Record the width of the stream at this point. (The width is the distance from one wetted edge to the other).

2. Use a stadia rod to measure the depth of the water in three places along the tape measure. One depth measurement should be at the midpoint of the stream; the other two should be at points equidistant from the midpoint to the wetted edges.

3. Use Data Sheet 5 to calculate the average depth of the starting point. Then calculate the cross sectional area of the starting point by multiplying the width by the average depth.

4. Repeat this process at two other points along the length of the channel used for the float trials. Use the finish line and a point in the middle of the stretch. Calculate the cross sectional area for these two other points.

5. Use Data sheet 5 to calculate the average cross sectional area for the whole channel used for the float trials.

Calculating Flow

Now it is possible to calculate the flow with the information you have gathered. Multiply the corrected average velocity by the average cross sectional area and record your flow value on Data Sheet 5.

A note on units: Most studies in the U.S. measure flow in **cubic feet per second (CFS)**. If you want to to compare your measurements easily with other studies, measure your depths in feet (cross sectional area in square feet) and your velocity in feet per second. This will give you flow in cubic feet per second. You'll want to use **cubic meters per second** in most countries outside the United States.

A Short Cut-Measuring Flow from a Pipe

Here is a short and simple way to measure flow from small streams or pipes. This method can be used when there are low flows and all the water can be directed into a container. Simply time how long it takes your container to fill. That is your flow.

If you're looking for a math exercise, you can use the following method to determine the volume of a bucket of unknown size:

The volume of a uniform bucket equals the area of the bottom multiplied by the height. The area of the bottom (for a round bucket) equals the radius squared multiplied by pi (3.14). The radius is the diameter divided by two.

volume = area x height

area = (pi)radius^2

radius = diameter/2

volume = (pi)radius^2 x height

Convert your radius and height values to feet before you calculate volume so your final volume value will be in cubic feet per second (see Calculating Flow, at left).

For those of you who are mathematically challenged, we suggest acquiring a bucket of known volume (5 gallons is common) or using a one gallon milk jug to determine your bucket's volume by counting the number of gallons needed to fill it to a given point. To convert gallons to cubic feet, divide the number of gallons by 7.48. (There are 7.48 gallons in one cubic foot.)

3. Position the bucket below the outfall of the culvert so that all of the water flowing out is caught by the bucket. Time how many seconds it takes for the bucket to fill completely with water.

4. The flow equals the volume of the bucket divided by the time it took to fill (flow = volume / time). Your answer should be in cubic feet per second.

TABLE 6: SWIMMING PERFORMANCE OF ADULT FISH

The table below outlines the "swimming performance" for various species of adult fish (from *Stream Enhancement Guide*, B.C. Ministry of the Environment)

Culverts should be designed so that normal flow velocities do not exceed the sustained speed capacities of migrating fish. Otherwise the culvert becomes a velocity barrier.

Cruising Speed = Speed fish can maintain for extended periods without fatigue.
Sustained speed = Speed fish can swim for short durations without fatigue.
Burst speed = Speed fish can swim only for a brief period before becoming fatigued.

Swimming speeds for average-sized adult fish of various species (metres/sec)			
Species	Cruising Speed	Sustained Speed	Burst Speed
Brown Trout	0 - 0.7	0.7 - 1.9	1.9 - 3.9
Carp	0 - 0.4	0.4 - 1.2	1.2 - 2.6
Chinook	0 - 2.7	2.7 - 3.3	3.3 - 6.8
Coho	0 - 2.7	2.7 - 3.2	3.2 - 6.6
Grayling	0 - 0.8	0.8 - 2.1	2.1 - 4.3
Lamprey	0 - 0.3	0.3 - 0.9	0.9 - 1.9
Shad	0 - 0.7	0.7 - 2.2	2.2 - 4.6
Sockeye	0 - 1.0	1.0 - 3.1	3.1 - 6.3
Steelhead	0 - 1.4	1.4 - 4.2	4.2 - 8.1
Suckers	0 - 0.4	0.4 - 1.6	1.6 - 3.1
Whitefish	0 - 0.4	0.4 - 1.3	1.3 - 2.7

U.S. Readers Note: This chart is in meters per second; you will need to convert to feet per second.

NOW WHAT?

Because it is often difficult to find a section of your stream that is straight and without turbulence, eddies, or backwater areas, your flow measurement will be an estimate. Make sure you are careful and consistent with your methods. Remember to not send the float down only the fastest path within the stream channel; this will skew your results. You need to estimate the range of velocities across the stream.

For a quality control check, ask a hydrologist to come out in the field with you and take velocity measurements with a flow meter, to see how your float method compares.

If you are concerned about migrating fish passage, compare your velocity results with the information provided in the box above. This will help you determine what species might find it difficult to swim through culverts and other potential velocity barriers.

Monitoring Storm Events

It is important to keep track of storm events when monitoring flow, because flow is so dependent on precipitation. Each time you measure flow, note how long it's been since it's rained and the approximate number of inches that fell in the last rainfall. If you can, monitor rainfall for at least one week prior to your flow measurement.

You can obtain precipitation and other weather information from your local weather service. However, to obtain data that is more specific to the microclimate of your stream, use your own rain gauge. They are inexpensive and easy to use, they just require regular attention. It is best if you can check your gauge every day when it rains.

Name_____ Group_____ Date_____ Time_____

Stream Name_____ Reach Name/#_____ Site Name_____

Site Location: Latitude___°___'___"N   River mile:_____
               Longitude___°___'___"W   Transect #:_____
Driving/Hiking directions:_____

Site Description: (e.g. transect #, landmarks, etc.)_____

Weather Conditions: ❏ Clear ❏ Cloudy ❏ Rain ❏ Other _____
                    Air Temperature: _____°___ (F or C)
Amount of precipitation in last storm_____ Time elapsed since last storm_____
Other recent weather information_____

Velocity Float Trials:

Trial #	Time	Distance
1		
2		
3		
4		
5		
6		
7		
8		
9		
10		
Total		

[ ] / = [ ]
total time/# of trials   avg. time

Cross Sectional Area:

Record depths at one-foot intervals Depth = D

Cross Section 1

#	D	#	D
1		11	
2		12	
3		13	
4		14	
5		15	
6		16	
7		17	
8		18	
9		19	
10		20	

Cross Section 2

#	D	#	D
1		11	
2		12	
3		13	
4		14	
5		15	
6		16	
7		17	
8		18	
9		19	
10		20	

Cross Section 3

#	D	#	D
1		11	
2		12	
3		13	
4		14	
5		15	
6		16	
7		17	
8		18	
9		19	
10		20	

([ ] + [ ] + [ ]) / 3 =
  sum   sum   sum

Average Cross Sectional Area = [ ] ft² (Multiply sums by 2 if depths were measured in 2 ft. intervals)

Average Surface Velocity = [ ] / [ ] = [ ] feet/sec.   X (0.8) = Average Corrected Velocity = [ ]
distance/avg. time   velocity correction factor

Flow = [ ] ft./sec. X [ ] ft.² = [ ] CFS
       Avg. Corrected   Avg. Cross   (cubic feet/second)
       Velocity   Sectional Area

Name_____ Group_____ Date_____ Time_____

Stream Name_____ Reach Name/#_____ Site Name_____

Site Location: Latitude___°___'___" N   River mile:_____
Longitude___°___'___" W   Transect #:_____

Site Description: (e.g. transect #, landmarks, etc.)_____
Driving/Hiking Directions_____

Weather Conditions: ☐ Clear ☐ Cloudy ☐ Rain ☐ Other _____
Air Temperature: _____°___ (C or F)
Amount of precipitation in last storm_____ Time elapsed since last storm_____
Other recent weather information_____

Velocity Float Trials:

trial #	time	Distance
1		
2		
3		
4		
5		
6		
7		
8		
9		
10		

___ / ___ = ___
sum/# of trials   avg time

Average Surface Velocity = [/] = _____
distance/avg. time   feet/sec.

Average Corrected Velocity = _____ X (0.8) = [ ]
feet/sec velocity correction factor feet/sec

Use Average Corrected Velocity to calculate Flow, using formula on page 3 bottom of following page.

Cross Sectional Area

Use Average Cross Sectional Area to calculate Flow, using formula on bottom of this page.

Cross Section 1

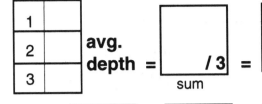

avg. depth = [ ] / 3 = [ ]
              sum   feet

Area₁ = [ ] x [ ] = [ ] ft²
      avg. depth   width

Cross Section 2

avg. depth = [ ] / 3 = [ ]
              sum   feet

Area₂ = [ ] x [ ] = [ ] ft²
      avg. depth   width

Cross Section 3

avg. depth = [ ] / 3 = [ ]
              sum   feet

Area₃ = [ ] x [ ] = [ ] ft²
      avg. depth   width

Average Cross Sectional Area =

[ ] Area₁

+

[ ] Area₂

+

[ ] /3
Area₃

= [ ] feet²

Flow = [ ] X [ ] = [ ] CFS
     Avg. Corrected   Avg.Cross   (cubic feet/second)
     Velocity   Sectional Area

CHAPTER SIX

The Spineless Ones

BENTHIC MACROINVERTEBRATES

TAKE HIM! TAKE HIM!

THE SPINELESS ONES

Macroinvertebrates are animals that do not have backbones, but are visible to the naked eye. **In this chapter, we use the terms macroinvertebrates and "bugs" interchangeably.** In streams, many "bugs" live on the stream bottom or on other substrates. These bottom dwellers are referred to as **benthic** macroinvertebrates. Many stream bottom "bugs" are insects, but many others are represented by freshwater aquatic worms, snails, clams, **crustaceans** (crayfish, crabs, shrimp, etc.), and **arachnids** (spiders and other 8-leggeds).

In this chapter, you will learn how to gauge the general health of your stream by using a sim-

plified field method to evaluate the "bug" population. You'll be able to use inexpensive or home-made equipment. The sorting and identification takes some time and practice, but is kept at a fairly general level.

The method is designed so you can conduct it at the stream site and release the "bugs" to the stream when done. It is **qualitative**, which means it does not analyze numbers of "bugs," but instead the presence or absence of different types. Nevertheless, it is detailed enough to diagnose the overall health of your stream.

This chapter also introduces a more **detailed lab method**, which requires "sacrificing" your specimens in order to conduct a more thorough and higher level analysis. It is a **semi-quantitative method**; it involves counting a portion of the "bugs" in your sample. For this method, you'll need a lab equipped with more expensive equipment, such as lighted magnifiers and microscopes. You'll also have to become familiar with some rather minute body parts, because this method requires identifying the "bugs" at a more detailed level. The reward is that you'll be able to detect more subtle impacts to your stream.

If you have little or no experience with "bugs," start out with the field method. If you have some experience, you may want to try the lab method, but read through the field method first. It introduces concepts and procedures you'll need for the more detailed analysis. You may find that some combination of the two procedures works best for you; feel free to adapt the methods as appropriate to your own situation.

Also, don't let a lack of resources stand in your way. If you are interested in the lab method but cannot afford microscopes, you may be able to recruit a school to help you that has an equipped lab.

"Bugs" as Indicators of Stream Health

Benthic macroinvertebrates have come to play key roles as biological indicators of stream health and water quality for the following reasons:

- "Bugs" represent important links in the food chain as recyclers of nutrients and food for fish.

- Because they are relatively sedentary residents of the stream bottom, benthic "bugs" often become a pollutant's captive

audience. Fish can (hopefully) swim away from some pollution problems. However, even regular chemical tests of the water column may fail to detect transitory events. And benthic "bugs," because they cannot swim away from pollution and can be affected by even subtle levels of degredation, are good indicators of stream health.

- There are many different types of stream "bugs." Each type has a specific set of requirements which the stream must provide for the organism to survive. Alterations to the stream may have a great impact on the abundance and distribution of different macroinvertebrate types.

- Some are intolerant of pollution. Their presence in the stream, like a "canary in a mineshaft," suggests healthy conditions. However, some "bugs" are quite tolerant of pollution. Taken together, the presence or absence of tolerant and intolerant types can indicate the overall health of the stream.

- Many "bugs" (especially those that are insects), tend to have short life cycles, usually one season or less in length. To assess the effect of a pollutant or flood on this year's population of juvenile salmonids, we would have to wait two, three or four years, depending on the species, to see a decreased number of returning adults. With "bugs" we wouldn't have to wait nearly as long to detect a problem.

- "Bugs" are easy to collect and the equipment used is fairly simple and inexpensive.

- "Bugs" are easier to identify than algae, which also have pollution tolerant and intolerant groups.

THE BODY SNATCHERS III

Classifying Macroinvertebrates (Bugs)

In order to use macroinvertebrates for assessing the health of a stream, you need to know how to identify them. Biologists categorize all forms of life into various levels of **taxonomic** groupings. The categorization is usually based on anatomical similarity and evolutionary relatedness.

Taxonomic levels, from most general to most specific, include **kingdom, phylum, class, order, family, genus**, and **species**. All animals are in the Kingdom Animalia. Insects and crustaceans are in the Phylum Arthropoda (critters with exoskeletons and segmented bodies and legs). Insects are in the Class Insecta or Hexapoda (six legs). To take a more specific example, black flies are in the Order Diptera ("true" flies) and the Family Simuliidae. Within this family, there are over 150 species in North America alone!

You may be picturing yourself in horror, trying to distinguish between the 150 species of black flies while they swarm around you biting ferociously. Fear not. If you use the simplified field method in this guide, you will only need to sort "bugs" into **major groups** (mostly orders).

Beyond the major group level, you will only need to distinguish one individual from another by noticing differences in body parts or overall **morphology** (body form and structure). For those of you who want more detail, the lab method involves identifying and sorting the "bugs" into different families. In neither case will you need to sort through the 150 different species of black flies!

BENTHIC MACROINVERTEBRATE (BUG) CLASSIFICATION

Kingdom

Animals
(Animalia)

Phylum

Segmented
Worm-like Creatures
(Annelida)

Segmented
Animals with Exoskeletons
(Arthropoda)

Clams, Snails, etc.
(Mollusca)

Class

Leeches (Hirudinea)

Segmented Worms
(Oligochaeta)

Crayfish,
Crabs, etc.
(Crustacea)

Insects
(Hexapoda
or Insecta)

Snails (Gastropoda)

Clams & Mussels
(Pelecypoda)

Order

Scuds
(Amphipoda)

Sowbugs
(Isopoda)

Crayfish
(Decapoda)

Mayflies (Ephemeroptera)

Stoneflies (Plecoptera)

Caddisflies (Tricoptera)

Crane Flies, Black Flies, Midges,
and other "True" Flies (Diptera)

Dragonflies, Damselflies (Odonata)

Dobsonflies, Alderflies, Fishflies (Megaloptera)

Beetles (Coleoptera)

Backswimmers, Water Boatmen, Water Striders,
and other "True Bugs" (Hemiptera)

SCUD STUD

Note: This is not a complete chart of all invertebrates; it contains only common
phyla, classes, and orders of stream benthic macroinvertebrates.

Basic "Bug" Morphology

Clams and snails are perhaps the easiest "bugs" to identify to the class level (Class Gastropoda for snails and Class Pelecypoda for clams) because of their well-known morphology. To identify something as an insect (Class Insecta), crustacean (Class Crustacea), or a worm (Class Oligochaeta) can be more difficult. The best clues to look for are the presence of a head and legs, the number of legs, and the presence of wings and an **exoskeleton** (an outer covering that provides support in lieu of an internal skeleton).

All adult insects have exactly six legs, while most crustaceans have more than six legs. Most adult insects have wings, while crustaceans do not. Both insects and crustaceans have heads and exoskeletons. Worms do not have any legs, heads, wings, or exoskeletons.

The insect body is segmented and divided into three major regions - head, thorax, and abdomen. The head appears to be a single segment but is actually composed of several fused segments. Just below the head, the **thorax** is composed of three distinct segments, with one pair of legs attached to each (hence a total of six legs). If an insect has wings, they are also attached to

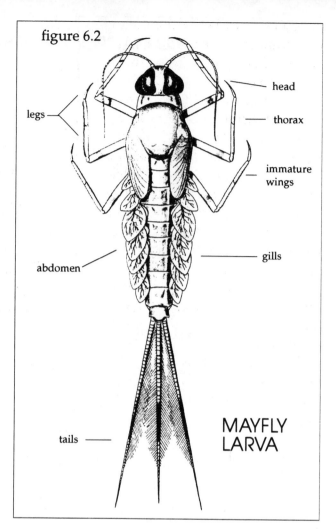

figure 6.2

MAYFLY LARVA

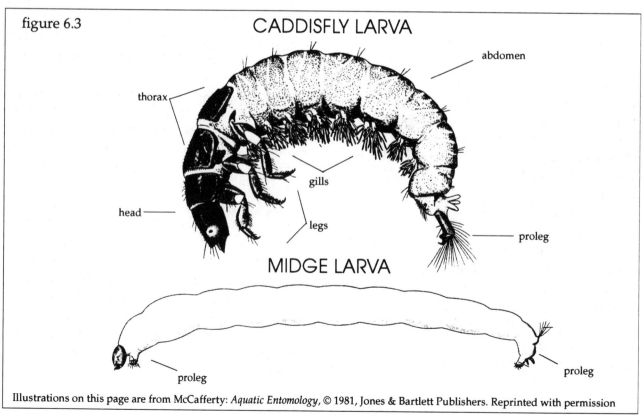

figure 6.3

CADDISFLY LARVA

MIDGE LARVA

Illustrations on this page are from McCafferty: *Aquatic Entomology*, © 1981, Jones & Bartlett Publishers. Reprinted with permission

the thorax. Just below the thorax, the **abdomen** is usually the longest region of an insect's body and is composed of several segments.

Unfortunately, aquatic insects can be difficult to identify as belonging to the Class Insecta because many are immature forms and thus are not fully developed. While many immature forms do have six legs, a distinct thorax and abdomen, and even undeveloped wings, many lack these features, and even appear headless.

In order to distinguish these immature insects from worms, look for other appendage-like attachments, such as gills, tails, filaments, and prolegs. **Prolegs** are fleshy, unsegmented, nubby leg-like structures attached to the thorax and/or abdomen of some immature insects. They are not to be confused with the regular legs of insects, which are segmented and usually longer and more slender. Some immature aquatic insects may have prolegs in addition to their six regular segmented legs.

Aquatic Insect Life Cycle

By now you've probably realized that in order to understand stream "bugs," a Streamkeeper needs to learn about the life cycles of aquatic insects, because many of our benthic buddies are immature, or **larval** forms of aquatic insects. Their adult forms are often winged, and can be seen flying alongside streams during the spring and summer.

The change from one form to another is known as **metamorphosis**. Some insects go through a complex set of changes known as **complete metamorphosis**. Insects such as flies, beetles and caddisflies begin their aquatic existence as an egg, laid in the water by the winged adult. The egg develops into a **larva**. The larva gradually transforms into a **pupa**. Pupae are generally non-moving and encased, like the cocoon of a butterfly or moth. The pupa undergoes dras-tic changes in anatomy and physiology, eventually emerging as a winged adult.

Other insects, such as dragonflies, stoneflies and mayflies, undergo a less complex set of changes known as **incomplete metamorphosis**. These insects also begin their lives in streams as an egg which metamorphose into a larva. The larva metamorphoses into a winged adult; there is no pupal stage (see figure 6.4).

The larval forms of insects which undergo incomplete metamorphosis are sometimes referred to as **nymphs**. In general, nymphs resemble their adult counterparts much more than the larval forms of insects which undergo complete metamorphosis.

As an immature insect develops, its exoskeleton does not grow like an internal skeleton would. Thus, a growing larva must shed or **molt** its exoskeleton and replace it with a new, larger one, to make room for its growing body. A larva during the interval between molts is referred to as an **instar** (figure 4.1). The number of instars an insect goes through to become an adult varies, even within a particular species, and can range from four to forty.

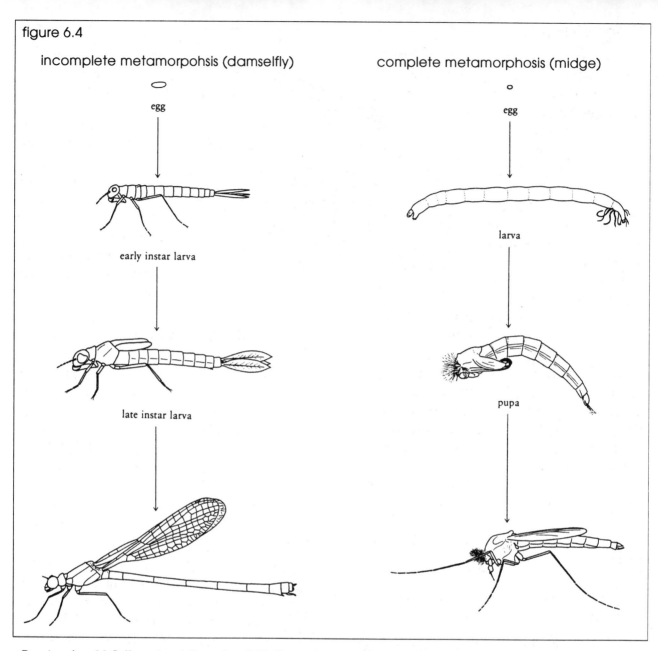

figure 6.4

incomplete metamorpohsis (damselfly)

egg

early instar larva

late instar larva

complete metamorphosis (midge)

egg

larva

pupa

Drawings from McCafferty: *Aquatic Entomology*, ©1981 Boston, Jones and Bartlett Publishers. Reprinted with permission.

Most aquatic insects spend the greater part of their lives as larvae. It is this larval stage which we most frequently encounter in our stream surveys. The larvae of some insects can remain in the water for more than a year, while others hatch into their adult forms after a shorter growth period.

The whole duration of aquatic insect life cycles ranges from less than two weeks to four or five years, depending on the species. For any particular species, life cycle events can also vary, depending on temperature, dissolved oxygen levels, day length, water availability, and other climatic and environmental conditions.

The adult lifespan of some aquatic insects, such as some species of mayflies, may be as brief as a few hours. This can be a hectic time for these ephemeral adults. They must quickly find another of the same species, mate, and deposit their eggs before they die to begin the cycle anew.

Remember that all macroinvertebrates are not insects. We must not forget our friendly aquatic worms, and freshwater snails, clams, and crustaceans. Most of these "bugs" spend their entire lives in water, from egg to adult, and do not undergo the same type of metamorphosis. Nevertheless, they share the habitats provided by the stream bottom with aquatic insect larvae.

"Bug" Habitats

This book has discussed different stream habitats such as riffles, runs, pools, and glides. Macroinvertebrates inhabit all of these. Typically, riffles where cobbles predominate with some boulders and gravel, contain the most diverse assemblages of "bugs." Because they often are the most rich in oxygen, riffles also tend to be home to those "bugs" that are sensitive to pollution.

Within runs, pools, and glides, there are a variety of **microhabitats** that provide food and shelter for macroinvertebrates, including rocky and gravel bottoms, finer sediments, living plants, **detritus** (litter, accumulations of leaves, twigs, bark, and other plant parts), and the surfaces of large boulders and woody debris.

There are also many critters, such as some species of freshwater mussels, burrowing mayflies, leeches, and aquatic worms, that prefer the more stagnant areas of fine sediments.

"Bugs" in the Stream Food Chain

Macroinvertebrates play an important role in stream food chains as the intermediate link between higher and lower feeding levels. They are food for **consumers** such as fish, birds, amphibians, reptiles, and other "bugs." For juvenile salmonids, resident trout, and many other species of fish, macroinvertebrates can be a principal diet item.

Some macroinvertebrates are **herbivores.** They feed upon plants or algae, which are known as **primary producers** because of their ability to use **photosynthesis** to produce their own food. Other "bugs" are **carnivores**, feeding on other macroinvertebrates, and even small fish and amphibians. Still others are **detritivores**, feeding on coarse or fine organic matter that falls into and is carried by their stream habitat. Macroinvertebrates in all feeding groups play an important role in recycling nutrients in the stream ecosystem.

Functional Feeding Groups

Macroinvertebrates do not always select food on the basis of whether it is plant, animal, or detritus. Many are more apt to select food simply because it fits the design of their food-gathering body parts. Macroinvertebrates are classified by their feeding habits into four **functional feeding groups**. They have developed a variety of adaptations which maximize the effectiveness of their preferred feeding strategy.

Shredders possess chewing mouthparts which allow them to feed on large pieces of decaying organic matter, such as leaves and twigs, which fall into the stream from trees and other plants in the riparian zone. Some have strong enough mouth parts to chew dead animal parts and/or living plant material when detritus is in short supply.

Scrapers or **grazers** remove attached algae from rock and wood surfaces in the current. They are found in areas where sunlight is able to reach the stream bottom, because without sunlight, algae cannot grow. Because these conditions often occur in larger, wider streams, many scrapers have developed adaptations for "hanging on" in relatively swift currents, such as flat, streamlined bodies or suction disks.

Collectors depend on fine particles of organic matter. **Filtering Collectors** are adapted for capturing these particles from flowing water.

Some caddisfly larvae spin nets for this purpose. Black fly larvae attach themselves to the substrate and filter particles using sticky hair-like fans.

Gathering Collectors gather small sediment deposits from the stream bottom or other substrates. Their mouth parts and appendages are designed for such activity, and many are adapted for burrowing into bottom sediments.

Predators consume other macroinvertebrates; they have behavioral and anatomical adaptations for capturing prey. Many have extensible mouthparts or raptorial forelegs adapted for grasping prey, and strong opposable mouthparts for biting and chewing. Some predators pierce their prey and suck body fluids with tubelike mouthparts. **YUM!**

The River Continuum Revisited

Recall the discussion on the river continuum in Chapter 1. The gradual changes that occur in a stream from headwaters to mouth affects the habitat structure and food base of the stream. As the habitats and food base change, so does the proportion of different functional feeding groups of "bugs:"

- **Shredders** tend to inhabit headwater streams and other areas with a high percentage of canopy cover. They play an important role in processing coarse organic matter into finer particles, which in turn can be used by other macroinvertebrates.

- **Scrapers** are more common in the middle reaches of a watershed where sunlight is able to reach the stream bottom, and thus algae is able to grow.

- **Collectors** tend to be common in all reaches, because fine particles are present in all stream types to some degree. However, they make up a greater proportion of the "bug" population in the lower reaches of a system where fine sediments tend to accumulate and the habitat is not suitable for shredders and scrapers.

- **Predators** are found in all habitat types. Because it takes many other "bugs" to supply their food, predators are usually found in small proportions relative to other feeding types.

figure 6.6
The River Continuum - Changes in the Macroinvertebrate Community as a Stream Widens.

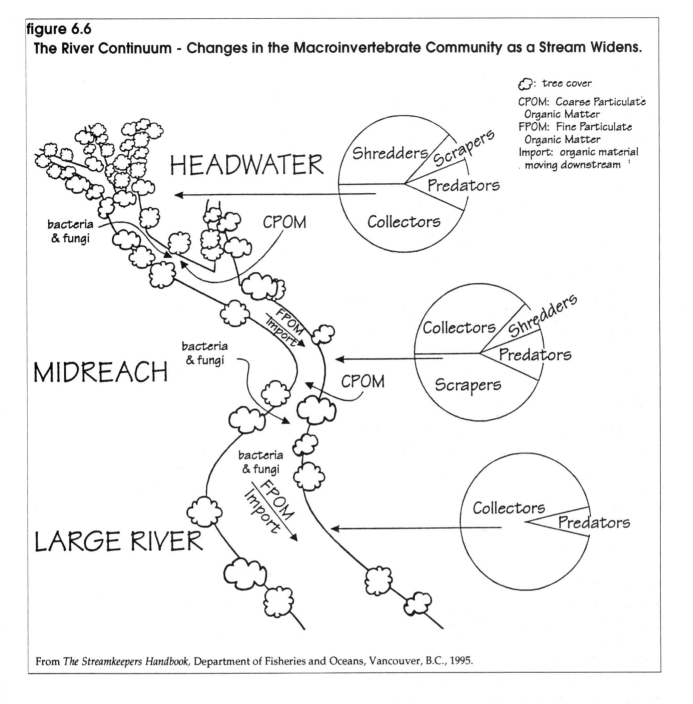

From *The Streamkeepers Handbook,* Department of Fisheries and Oceans, Vancouver, B.C., 1995.

Breathing Adaptations

The art of breathing in an aquatic environment poses a great challenge to macroinvertebrates, because oxygen in water exists as **dissolved oxygen**, which is less abundant than the atmospheric oxygen available to terrestrial beings. To utilize dissolved oxygen in their underwater home, stream "bugs" have developed countless anatomical and behavioral strategies:

• Most have soft or membranous areas of the body wall through which oxygen diffuses.

• Many have external **gills**, or membranous outgrowths of the body wall, which increase the surface area for oxygen uptake. These gills can be platelike, filamentous, tubelike, or fleshy.

• Some ventilate parts of their bodies to increase their available oxygen supply. Certain mayfly larvae beat their abdominal gills;

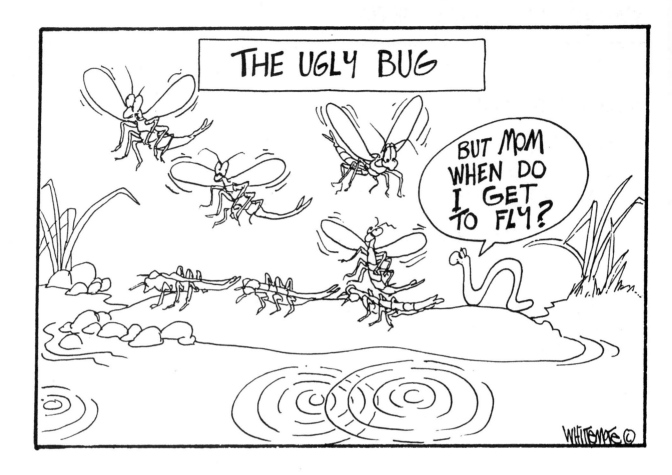

some caddisfly larvae are known for their undulating abdominal movements.

- Some possess functional **spiracles**, openings on the outer body wall which lead directly to a network of tubes that distribute oxygen throughout the body. These spiracles, however, must not be in direct contact with the water or the organism will drown.

- The spiracles of some macroinvertebrates are covered and kept dry by a **plastron**, a thin film of air held by tiny hairs. The plastron looks like an air bubble through which dissolved oxygen can diffuse. Examples of organisms which exhibit this strategy include riffle beetle adults and some midge and crane fly larvae.

- Other aquatic macroinvertebrates with spiracles maintain some bodily contact with the air-water interface via tubelike structures and actually breathe atmospheric oxygen. Mosquito pupae and water scorpions are examples of such organisms.

- Others who breathe atmospheric oxygen through spiracles are able to carry their own air supply under water via a plastron or bubble held by wings or other hair-covered body parts. Many adult water beetles and aquatic true bugs utilize this strategy.

Creatures which breath atmospheric oxygen are generally not benthic; rather, they are free swimming or float in the waterstream. They are also able to inhabit areas with low dissolved oxygen levels.

FIELD PROCEDURE: MACROINVERTEBRATE SURVEY

EQUIPMENT NEEDED

Collection

sampling net (see Table 8, page 137)

sieve bucket and/or sieve (see figure 6-8 page 138)

small buckets or quart-sized containers

waterproof boots/waders

waterproof, insulated, elbow-length gloves (if working in polluted or cold water)

Sorting & Identification

large shallow white trays (9" x 13" or larger; 1-3" deep)

small, shallow white plastic or styrofoam containers (meat trays and cream cheese, pudding and tofu containers, etc.)

2-4 white ice cube trays, pre-labeled

"bug" handling "tools":

- tweezers (for hard-bodied "bugs" that don't squash easily)
- white plastic spoons (for softer, more fragile "bugs")
- clear plastic pipets (for small & fast "bugs")
- small paint brushes (well suited for mayflies)

hand lenses

Data Sheet 6A, "Macroinvertebrate Survey (Field Method)"

macroinvertebrate keys (see pages 148-153)

pencils

clipboard

Pre-labeling Ice Cube Trays

Along the long edge of your ice cube trays, use labeling tape and waterproof markers to label each row of 2 cubes. Label the rows with the following major taxonomic groups:

- mayflies
- stoneflies
- caddisflies
- dobsonflies/alderflies/fishflies
- dragonflies/damselflies
- true flies
- beetles
- crustaceans
- snails/clams
- worms/leeches
- others

As you become familiar with the "bugs" in your stream, you may find there is a more appropriate way to label your ice cube trays. for example, if you never find any dragonflies or damselflies, you can stop labeling a row for that order. Or, if you tend to find several different taxa of mayflies, you can set aside 3 or 4 rows labeled with that order, so you can keep each taxa in a separate "cube." You may find that you need 2-4 ice cube trays to best accomodate your particular "bug" community.

WHAT?

This simplified field method will help you evaluate the general health of your stream by determining the presence and absence of different types of macroinvertebrates. This method is designed so that you can sort and identify the "bugs" in the field at your collection site, and return them to the stream when finished. You will need to identify them by **major groups** (mostly orders), and sort them more specifically simply by noticing morphological differences.

WHERE?

This procedure focuses on riffle habitats because in general they support the most diverse assemblages of "bugs." Riffles are most frequently sampled and compared in studies of stream macroinvertebrates and tend to be easier to sample than other habitats.

Ideal riffle characteristics for this procedure follow:

- a current velocity between 0.4 and 2.0 feet per second *
- a depth between 0.5 and 2.0 feet
- a predominance of cobbles (>50% if possible)
- at least 3 x 3 square feet in size if possible
- undisturbed and safe for you to wade in

* Use the velocity float method on page 110 to make sure the riffles you choose are within this range. You do not need to do 10 trials and average; just one or two trials will do.

For each riffle you choose, collect your sample at its upstream end, just below a pool. This area is called the **head** of the riffle. The head of a riffle is often the most productive; the "bugs" tend to migrate to these areas because they offer a competitive feeding advantage.

While riffles typically support the most diverse "bug" communities, some species prefer other stream habitats, such as silty bottoms, aquatic plants, **leaf packs** (clumps of fallen leaves and other organic debris), and large boulders and woody debris. If you only sample riffles, you may miss the species that prefer these other areas. Moreover, if your stream lacks sufficent riffle areas, you will have to sample these other habitats to find the "bugs." Methods for sampling in these other areas are outlined briefly in the field procedure on page 132.

How many "bug" collection sites do you need? We recommend you start out with one good riffle site per stream reach that you monitor. As you expand your macroinvertebrate surveying, you may need to scope out additional sites to find other types of "bug" habitats. If you find one area that provides a variety of habitat types, including a good riffle, you may want to stick with that site.

Mark your collection sites in the field and document their locations on Data Sheet 6A and in your Quality Assurance Plan.

WHEN?

Because of their varied lifespans and responsiveness to environmental conditions, macroinvertebrate populations fluctuate with the seasons. Therefore, we recommend you collect "bugs" four times a year, once each season. If you are unable to collect four times a year, twice is adequate.

The most important collection time is early autumn before leaf fall. Populations are more stable at this time because fewer adults are emerging from the water. This early fall is also a good time because it often represents a "worst case scenario" after a summer of high temperatures and low flows.

The second most important collection time is in the spring, after snowmelt and before the leaves are fully out. If you compare a sample at this time with an autumn sample, you can evaluate how the community changes over the summer.

HOW?

Collecting the "Bugs"

Riffles

1. Approach the riffle from downstream and place the net at the downstream edge of the area you wish to sample. The net should be perpendicular to the flow and lay firmly on the stream bottom. (If using a kick net, tilt it back at a slight downstream angle, but not so much that water flows over the top of the net. Anchor the bottom with cobbles to prevent "bugs" from escaping).

2. Sample organisms from a total stream bottom area of approximately one square meter. If you are using a kick net that is one meter wide, you can sample the 3 x 3 square foot area just upstream from the net. If you are using a D-net or surber, repeat steps 3 and 4 in different areas

until you have sampled a total of approximately 3 x 3 square feet of the stream bottom.

3. Pick up rocks from the sample area in front of the net that are over two inches in diameter. Hold them in front of the net and below the water surface. Gently but thoroughly rub the organisms from the rock surfaces so they flow into the net. Place the "clean" rocks outside the sample area after the organisms have been removed. Continue until all rocks in the sample area have been rubbed.

Note, Miss Mayfly says: "Be wary of sharp rocks, and especially, glass or metal objects." If you suspect such litter in your stream, do not carry out this step without protective gloves. Also, make sure you use gloves if you suspect that your stream is highly contaminated. Wash your hands when you return from the field.

4. After rocks and debris have been rubbed, step inside the sample area. Starting from the upstream side and working your way to the front of the net, disturb the stream bed by kicking or using a garden fork. Do this until the substrate is thoroughly disturbed in the entire area, digging down as far as possible. (You may want to sing a tune and do a "bug bugaloo" during this step. Collecting "bugs" can be fun as well as scientific).

5. Remove the net with a forward scooping motion. (If using a kick net, be careful not to let water flow over the top or bottom of the net. Fold the net in half so the side handles are together.) Carry the net to a comfortable spot on the edge of the stream.

6. Gently scrape the "bugs" into one corner of the net. Knock the bulk of your sample from the net into a sieve bucket. Wash the rest into the bucket by pouring water from the smaller buckets over the back of the net. Make sure the entire sample of "bugs" is in the sieve bucket.

7. Place the sieve bucket in the stream and let if fill partially. Swirl the bucket to "wash" silt out of the sample. If there are large pieces of organic debris (leaves, sticks, etc.), "wash" them free of "bugs" so the "bugs" remain in the bucket. Then remove the debris from the sample.

Bottom Sediments and Aquatic Plants

Use a long-handled net to collect samples of bottom "muck" and sweep through stands of aquatic plants. Often these habitats are found along the bank edge in backwater areas. Transfer contents from the net into the sieve bucket and sift silt from the sample.

Leaf Packs

Clumps of leaves and other debris are caught by large rocks and logs. The most productive clumps consist of older materials that have begun to decay. Simply pick up the clumps and place them into a sieve bucket. Wash materials so the "bugs" remain in the bucket. Then remove the debris from the sample.

Large boulders and woody debris

Sampling this type of "bug" habitat requires visually inspecting all exposed surfaces and picking the "bugs" off by hand. Inspect boulders, logs, stumps, etc. that occur in all types of stream habitats (riffles, pools, runs, and glides).

Sorting & Identifying the "Bugs"

1. Transfer your sample from the sieve bucket to a large, shallow white tray. Spread the sample over the bottom of the tray into a thin layer of "bugs." Add just enough stream water to cover the sample. If your sample is too large to spread into a thin layer, use only a portion of the sample at a time, or better, use more shallow white trays and get other Streamkeepers to help you.

2. Fill your labeled ice cube trays with stream water. Line them up end to end so the labeled edges face you. Use your "bug" handling "tools" to sort organisms into the ice cube trays. Small white plastic containers or styrofoam trays work well to hold "bugs" while you use hand lenses to examine and identify them.

3. Use the **dichotomous key** on pages 148-153 to separate the "bugs" into their major groups as labeled on the ice cube trays. A dichotomous key guides you through a series of decisions about the morphology of a "bug" in order to arrive at its appropriate taxonomic grouping.

You will not need to identify the "bugs" beyond the major groups labeled on the ice cube trays. However, within each major group, try to distinguish between individuals that represent more specific groups.

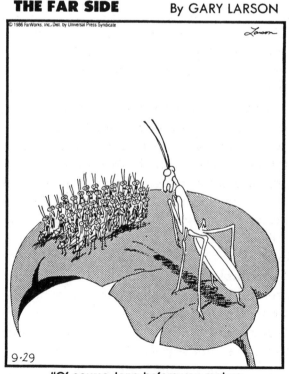

THE FAR SIDE   By GARY LARSON

"Of course, long before you mature, most of you will be eaten."

figure 6-7

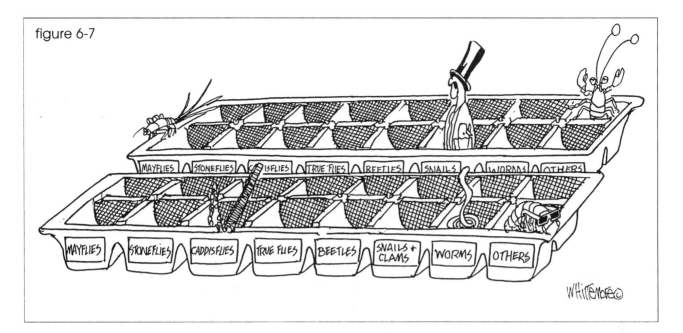

For example, you may find in a sample of seven mayflies that three individuals look identical but distinct from the four others, and these four themselves look identical. This may mean that the 7 individuals represent two distinct families within the Order Ephemeroptera (mayflies).

The most specific grouping that you can distinguish is referred to as a **taxon.** A taxon is simply a general term for a taxonomic group of organisms at a particular level (**taxa** if used in the plural). In some cases, the taxa you distinguish will be different species or genera; in some cases they may be different families; it all depends on the level of detail that you can observe.

Be aware that size and coloration alone are not valid criteria for distinguishing different taxa. While some species are definitely smaller than others, individuals within a species may vary in size depending on their age. Take care not to count different instars of the same species as distinct taxa.

More appropriate characteristics to use in distinguishing among taxa include the shape of the body and head, and the location, structure, and length of the gills, tails, eyes, mouthparts, antennae, and prolegs. The picture key on pages 154-161 will help you distinguish between the different family level taxa of the mayfly, stonefly, and caddisfly orders.

You may find many different taxa within each major group labeled on your ice cube trays. Things stay less confusing if you store representatives from each taxon in separate "cubes." If you run out of cubes for a major group, use additional ice cube trays and line them up, long sides together, behind your labeled trays. Or you can re-label the trays, assigning more than one row of cubes to the more abundant major groups.

Unless your sample is small (less than 100 individuals), you will not be able to pick out every last "bug." This field method is designed to examine the number of different taxa, not to count the total number of "bugs" (although you will be asked to do a gross estimate of numbers in each major group).

To be most efficient with your field time, concentrate on picking only a few representatives of each taxon out of the sample. However, be thorough as you search through the sample, looking for different taxa. Try not to let any go unrepresented. Some "bugs" are very small, so look carefully with your hand lense.

4. Use Data Sheet 6A to record the number of distinct taxa that fall within each major group. Use the dichotomous key or the mayfly/stonefly/caddisfly picture key to determine the feeding strategy of each taxon. For each major group, record the number of taxa that are shredders,

WHAT HAPPENED TO DADA?

HE'S DATA NOW.

scrapers, gathering collectors, filtering collectors and predators.

5. There may be some "bugs" that you have difficulty placing into a major group. You can preserve them in a 70% ethyl alcohol solution (a mixture of seven parts ethyl alcohol and three parts water). Bring your preserved specimens to a "bug" expert who can help you identify them. You may also want to preserve representative specimens of each known taxon in separately labeled vials.

Having such a "bug" library on hand will help you with further macroinvertebrate sorting sessions. After you have recorded your findings on the data sheet and preserved representative "bugs" as needed, return your remaining live specimens gently to the stream.

6. If you are able to pick every last individual "bug" out of the sample, you can take your analysis one step further. You can find a value for **sample density**, the total number of individual "bugs" in the sample. You can also find the density of each taxon, major group, and functional feeding group. With all this information, you can use most of the analyses found in the more detailed lab method. Refer to the section entitled "Lab Procedure, Macroinvertebrate Survey" on page 140.

7. (For this step, you can retreat indoors if you like). Add up the columns on Data Sheet 6A to find the total number of taxa, the number of **EPT** taxa, and the number of taxa in each functional feeding group. EPT refers to mayflies (Order **E**phemeroptera), stoneflies (Order **P**lecoptera), and caddisflies (Order **T**ricoptera).

NOW WHAT?

Analyzing Your Results

Taxa Richness

The taxa richness, or total number of taxa, tells you important information about the diversity of the "bug" population in your stream. Knowing that there are caddisflies in your stream is good information, but knowing there are three different families (taxa) of caddisflies is even better information. The greater the taxa richness, the higher the diversity of your "bug" population.

In general, streams with a higher diversity of "bugs" are considered healthier than those with a lower diversity. Pollution often causes a decline in diversity by favoring a fewer number of pollution tolerant taxa that are able to outcompete the more sensitive ones.

However, a moderate amount of **organic pollution** (excess nutrients from leaky septic tanks, fertilizers, etc.) sometimes causes an increase in taxa richness, especially in high altitude streams that are naturally low in taxa and number of individuals.

Biologists have determined the pollution tolerance of many common macroinvertebrates. Usually, in discussions on "bug" pollution tolerance, the pollution in question is excess nutrients or sediments, causing low dissolved oxygen conditions. In general, mayflies, stoneflies, and caddisflies have the lowest tolerance to pollution, while midges, aquatic worms, leeches and blackflies have the highest tolerance. Beetles, craneflies, and crustaceans tend to be in the middle ("somewhat tolerant").

Of course, the life of "bugs" is not so simple. Because there are so many species within each major group, there are also many variations in pollution tolerance within each major group. Some caddisflies are very tolerant, while some midges are very sensitive. Because of this variation, a pollution tolerance analysis is best left for the more detailed lab method, which involves family level identification.

EPT Richness

The EPT richness, or number of mayfly, stonefly, and caddisfly taxa, provides important information about your stream because these orders are, in general, some of the most sensitive to pollution. The EPT richness usually declines with pollution (although some mayflies and caddisflies are moderately pollution tolerant).

Many species of midges, black flies, crustaceans, aquatic worms, leeches, and snails are more tolerant of pollution, and thus they tend to move into niches vacated by mayflies, stoneflies and caddisflies when areas become polluted. This shift may simplify and destabilize the structure of the "bug" community, and thus reduce the biotic integrity of the stream ecosystem.

In general, between 8-12 EPT is considered good, but some naturally unproductive high altitude streams may have lower EPT richness and still be in a pristine state.

Number of Taxa in each Functional Feeding Group

In general, a healthy stream should support a variety of functional feeding groups. Certain human impacts may create an overabundance of a particular food source, which could lead to the predominance of a particular feeding type and an unbalanced biotic community. Recall, however, that the natural variation of the river continuum will influence the relative proportion of feeding groups present.

The nature of your sampling sites will also affect the diversity of feeding groups. The more various the types of habitats you sample, the greater the diversity of feeding groups you should find.

Seasonal changes will also affect the presence of feeding groups. Shredders tend to be more abundant in the fall when organic litter is most plentiful. In streams with canopy cover, scrapers will likely be present in the winter but absent during the summer when leaves shade the stream, limiting algae growth.

Drawing Conclusions

Table 7, adapted from the Canada Department of Fisheries and Oceans, British Columbia Salmonid Enhancement Program, can help you figure out what your "bug" results might mean for your stream.

It may be difficult to attach a formal rating to the health of your stream (e.g. "good," "fair," etc.) based on the taxa richness, EPT richness, and the relative proportions of feeding groups present. The relationship between these results and the health of your stream depends on the character of your particular stream.

As you know, the macroinvertebrate community is affected by many factors, such as the condition of the stream bottom, the depth and velocity of the water in the stream, day length, temperature, and water quality factors such as pH, dissolved oxygen, organic nutrients, bacteria, and toxics. As you collect "bug" data in combination with your stream reach survey, stream bottom analysis, flow measurement, and water quality tests, you will become more

TABLE 7 ANALYZING "BUG" DATA

Observation	Analysis
• high diversity, lots of stoneflies, mayflies, and caddisflies	• no problem, good water quality
• low diversity, high density, lots of scrapers and collectors	• organic pollution (nutrient enrichment) or sedimentation; lots of algal growth resulting from nutrient enrichment
• only 1 or 2 taxa, high number of collectors	• severe organic pollution or sedimentation
• low diversity, low density, or no "bugs," but the stream appears clean	• toxic pollution (e.g. chlorine, acids, heavy metals, oil, pesticides), or naturally unproductive due to limited light or nutrients (small high altitude streams)

familiar with the character of your stream and how different parameters affect its "bug" population.

Probably your biggest challenge will be to sort out the natural variation of your stream from human impacts. If you are doing a macroinvertebrate survey to assess the impact of human activity on your stream, you need to know what you would expect to find in your stream under natural conditions and at different times of the year. See Chapter 3 for more information about how to design your stream monitoring program to sort out natural from human-caused variation.

As you survey your stream's "bug" community over time, make sure that your methods are consistent. Be aware that taxa and EPT richness will increase with increasing sample size (total number of bugs collected). With this qualitative field method, it is difficult to factor out the effect of sample size on your results, because you are not counting the number of "bugs" in your sample.

However, you can ensure that your sample areas are always the same size. If you sample riffles, stick to the 3 x 3 square foot area. If you sample other habitats, try to standardize your

sample size as best you can, e.g., pick up a set number of leaf packs of a given size; make a set number of "jabs" with your long-handled net; and sample boulders and large woody debris until you pick a set number of "bugs."

There are other "bug" **metrics** (values or measurements calculated from raw data) that can better attach a formal rating to your stream. However, these metrics can only be used with a more quantitative method in which you preserve the "bugs" so you can count and identify them in more detail. If you are interested in such further analyses, see the section entitled "Lab Procedure, Macroinvertebrate Survey" on page 140.

While a more detailed method can provide a more sensitive analysis of your stream's "bug" community, rest assured that the simplified field method still provides invaluable information. In fact, that data may be the only macroinvertebrate information available on your stream. While more government agencies are recognizing the importance of "bugs" in determining stream health, many have yet to develop a macroinvertebrate monitoring program.

TABLE 8 "BUG" COLLECTING EQUIPMENT

There are a variety of nets used for collecting "bugs." A few different types are listed below, along with their pros and cons. Choose the net that best suits your needs. No matter what type you choose, use 0.6 mm mesh size, the standard recommended by the U.S. Environmental Protection Agency. Larger mesh may allow some "bugs" to escape; smaller mesh may clog with sediment and other debris. See Appendix B for instructions on how to make the nets, or see the "Equipment Sources" section in Appendix C for purchasing information.

Nets Used in Moving Water	Type & Description	(+) (-) Pros & Cons
	**Kick Net** Square mesh screen attached to two dowels; usually a little larger than 3 x 3 square feet in size	+ can collect good sample all at once + easy to make + can use in deeper water than others - requires 2 people to hold net down in current - a bit unwieldy when transfering sample into holding container
	**D-Net** Mesh screen attached to "D" shaped frame at end of a long handle; flat edge of the "D" is placed on the stream bottom; usually no more than 1.5 feet wide	+ only requires 1 person to hold in current + easier to transfer sample into holding container - too small to collect 3 x 3 square foot sample all at once (must use composite samples) - some "bugs" may drift over top of net in deeper water - a bit more difficult to make your own
	**Surber Sampler** Mesh screen attached to one of two square frames that are hinged together; second frame used to outline the sampling plot on the stream bottom; usually one square foot in area	same pros and cons as D-Net, except: + can be more precise when collecting (can sample an exact area of the stream bottom) - even more difficult than D-Net to make

Table adapted with permission from the TVA

	Use & Description	Pros and Cons
Dip Net 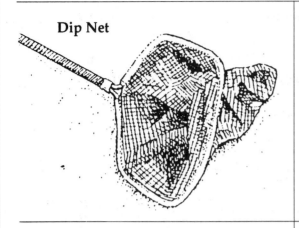	Mesh screen attached to round, oval, or triangular frame at end of a long handle (for use in slow-moving backwaters & pools - aquatic vegetation and silty bottom samples)	+ long handle allows samples to be taken without getting your feet wet. Can also be used to catch fish.
Sieve 	Square or circular frame with a 0.6 mm screen or #30 mesh bottom (for cleaning silt from newly collected samples in the field; also for washing preservative and additional silt from samples in the lab, just before analysis)	+ easier to use in lab than sieve bucket for washing samples before analysis - more difficult to use in field than sieve bucket for cleaning silt from samples
Sieve Bucket 	A bucket with a sieve bottom as described above (for cleaning silt from newly collected samples in the field)	+ more unwieldy to use in lab for washing samples before analysis - easier to use in field for cleaning silt from samples (holds more volume, thus easier not to lose any of sample)

DATA SHEET 6A MACROINVERTEBRATE SURVEY (FIELD METHOD)

Name_____Date_____Time_____

Stream Name_____Reach Name/#_____Site Name/#_____

Driving/Hiking Directions_____

Weather Conditons: ___Clear   ___Cloudy   ___Rain   ___Other_____

    Air Temperature_____°___   Water Temperature_____°___

    Recent Weather Trends_____

major group	# taxa in each Feeding Group*					# taxa
	SH	SC	GC	FC	P	
mayflies						
stoneflies						
caddisflies						
true flies: midges						
craneflies						
blackflies						
others						
dobsonflies, alderflies, fishflies						
beetles						
dragonflies/damselflies						
crustaceans: crayfish						
sowbugs						
scuds						
others						
snails						
clams & mussels						
worms & leeches						
others						
Totals →						

Total # mayfly, stonefly, caddisfly taxa

→ EPT Richness

Taxa in each functional feeding group

↑ Taxa Richness (total # of taxa)

Site Description:
Habitat Type_____
_____

Avg. depth_____

Avg. current velocity
_____

Other Comments_____
_____
_____

Site Location (lat/long. transect # and/or

rivermile_____

*Major Feeding Groups

SH - Shredder
SC - Scraper
GC - Gatherer Collector
FC - Filter Collector
P - Predator

Lab Procedure: Macroinvertebrate Survey

EQUIPMENT NEEDED

You'll need the same equipment as listed in "Field Procedure: Macroinvertebrate Survey" on page 129, with the following additions:

- quart-size wide-mouth jars with water tight lids (one per replicate sample)
- 70% ethyl alcohol (a solution of 7 parts 100% ethyl alcohol to 3 parts water)
- labeling tape
- wax pencils or waterproof markers
- 0.5 mm or #30 mesh sieve (or use your sieve bucket)
- large shallow white trays (9" x 13" or larger, 1" deep)
- lighted magnifiers
- dissecting scopes, at least 40x
- four compartment petri dishes (instead of ice cube trays)
- small vials with water tight lids
- macroinvertebrate keys
- Data Sheet 6B "Macroinvertebrate Survey (Lab Method)"

WHAT?

This more detailed method will help you evaluate the health of your stream using a semi-quantitative macroinvertebrate analysis. The method involves collecting 3 replicate samples per site. It also requires a lab so you can count a random subsample (about 100 individuals) for each replicate and identify the "bugs" to the family level.

This level of analysis will allow you to detect more subtle impacts to your stream's "bug" community. By counting the "bugs" and identifying them to the family level, you can calculate various **metrics** (values or measurements calculated from raw data) that can help you determine a specific rating for the health of your stream.

WHERE?

Use the same types of sampling locations as described in "Field Procedure: Macroinvertebrate Survey," page 131.

WHEN?

Sample at the same times as described in "Field Procedure: Macroinvertebrate Survey."

HOW?

Collecting the "Bugs"

This procedure involves collecting three **replicate** samples of "bugs." Replicates are multiple samples taken from the same site at the same time. Average values for a site are calculated by averaging the replicates. Using the average of three replicate samples provides more quality assurance for your data than just analyzing one sample.

Follow the same procedure as described in "Field Procedure: Macroinvertebrate Survey," to collect three replicate samples from one or more riffles at your "bug" sampling site. Each replicate should be collected from a one square meter area. If you do not have a riffle that is large enough (three square meters), use more than one riffle. Try to make sure, however, that all replicates are collected from the same general site within your 500' long reach. Keep the replicates separate, sorting and identifying the "bugs" in each, using the method outlined below.

If you are collecting "bugs" from other habitats besides riffles, collect three replicates for each habitat type. Do not treat the different habitat types as different replicates, i.e., do not use a riffle sample as replicate 1, an aquatic plant sample as replicate 2, and a large woody debris sample as replicate 3. Averaging the values from the different habitat types is like averaging apples and oranges.

After collecting them, gently scrape each replicate sample of "bugs" from your sieve bucket and place into a sample jar. Pour 70% ethyl alcohol into the jar to cover the sample and top it with water by one inch. Cap the jar tightly and label with the date, site, habitat type, and replicate number. Bring the jars back to your lab for analysis.

Sorting & Identifying the "Bugs"

Follow the procedure below for each replicate sample:

1. Rinse the sample in your sieve bucket or some other sieve (0.6 mm or #30 mesh) to wash off the alcohol.

2. Mark a grid of 12 squares on the bottom of a shallow white tray with a wax pencil or waterproof marker. This grid will allow you to estimate the total **density** of your sample (the total number of individuals) by picking a sub sample of about 100 "bugs." Make sure the squares are equal in area. Spread the sample into the tray, making sure all "bugs" are transferred. Cover the bottom of the tray with about 1/4" of water. Spread the "bugs" out over the tray so they are evenly distributed, all the way to the corners of the tray.

3. You will pick one square at a time until you reach 100 "bugs" total, picking every individual out of each square before moving to the next one. Each time you need to select a square, roll a pair of dice so your choice is random.
Set up your lighted magnifier over the tray to make sure you find all the small individuals as well as the large ones. Check under leaves, rocks, and within clumps of algae and other organic matter. As you pick the "bugs" out of the tray, sort them into major groups using the dichotomous key on pages 148-153. Place them, by major group, into petri dishes.

4. Continue this process until you have picked 3 squares. If you have picked at least 100 individuals, you can stop. If not, continue choosing and picking squares until you have reached 100 "bugs." Finish picking all the individuals from the last square that you happen to be on when you reach 100.

Don't worry if you have to pick the entire tray to reach 100 "bugs" or if you've picked the entire tray and haven't reached 100. This just means that your sample is small. Make sure, however, that you are not missing any "bugs"; some are very tiny and can be difficult to see.

5. For each replicate, record the total number of "bugs" that you picked on page 3 of the data sheet (in the chart under Part 3: "Projected Average Density for the Sample"). Also record the number of squares that you picked for each replicate. You will use these values later when you calculate the projected density of the sample.

6. Identify and sort the "bugs" into family groups using the Dichotomous Key on pages 148-153 and the Picture Key on pages 154-161.

7. Make sure you label the petri dishes with the sample site, habitat type, and replicate number before moving on to the next replicate. You may find it helpful to label the dishes with major group and/or family names as well.

8. For each replicate, record the density (number of individuals) for each family on Data Sheet 6B. Use the dichotomous key and the picture key to determine the functional feeding group for each family and record this information on the data sheet.

You may find some families that have individuals from different feeding groups. In this case, try to note this information as best you can on your data sheet. For example, most craneflies are in the same family (Tipulidae) but some are shredders and some are predators. If you had four shredder cranefly and one predator cranefly in your sample, you could write "4-S / 1-P" in the feeding group column for that family.

9. Use Table 9 on pages 162-163 to determine the pollution tolerance value for each family and record this information on the data sheet.

10. When you are finished, save your sample in small labeled vials. This is a good quality control procedure because it allows your results to be verified if necessary by others. Use some representative specimens from each family to start a "bug library." This can be an excellent resource for future sorting and identifying work.

Editor's note: Hanford, Washington, is located near the Columbia River between Washington state and Oregon. It is the former site of a nuclear weapons facility, and presently the site of a nuclear power plant and nuclear waste disposal facility. It is a major EPA Superfund clean-up location.

NOW WHAT?

Analyzing Your Results

Taxa Richness, EPT Richness

Count the total number of taxa represented in the whole sample and record this value on Data Sheet 6B. Then count the total number of EPT taxa (mayflies - Order **E**phemeroptera, stoneflies - Order **P**lecoptera, and caddisflies - Order **T**ricoptera). To evaluate your results, see the discussion on taxa and EPT richness on page 135 in the "Field Procedure: Macroinvertebrate Survey."

For taxa and EPT richness, the main difference between this lab procedure and the simplified field method is that values are controlled for the total number of "bugs" in the sample (density). This is important because richness increases with increasing sample size. Standardizing density to approximately 100 individuals allows you to compare taxa and EPT richness values among different samples without worrying about the effect of sample size on these values.

Projected Density of the Sample

This lab method allows you to project a value, based on your subsample, for the actual density of the sample. On page 3 of the data sheet, you have already recorded the total number of "bugs" and squares picked for each replicate. Calculate the projected density for each replicate as follows:

Projected replicate density =

$$\left(\frac{\text{Total bugs picked}}{\text{Total \# squares picked}} \right) \text{ x 12 squares}$$

For example, if you picked 108 "bugs" out of a total of 3 squares, the average density of each square would be 108/3 = 36 "bugs"/square. Assuming your tray had 12 squares, the projected density of the whole replicate would be 36 X 12 = 432 "bugs." If you marked your tray into a different number than 12 squares, make sure you adjust for this on the data sheet by using the appropriate number in your calculations.

Calculate the average projected density of the whole sample by averaging the projected values of the three replicates. Record your result on Data Sheet 6B.

A stream with a higher "bug" density per sample is not necessarily healthier than a stream with a lower "bug" density per sample. Organic pollution often increases the density by adding additional nutrients to the water. However, toxics and physical stress (sedimentation) will usually decrease the density.

Percent Composition of the Major Groups / Dominance

On the data sheet, calculate the average density for each family. Then sum these values within each major group to find the total average density for each major group. Finally, sum the values of all the major groups to calculate the total average density of the sample. Record this last value at the bottom of page 2 on Data Sheet 6B. (Note that you can also calculate the total average density of the sample by averaging the total number of "bugs" picked over the three replicates.)

Calculate the percent composition of each major group by dividing the total average density for each group by the total average density of the sample. (Do not use the projected density for this calculation.) Record your results on page 3 of the data sheet.

This information helps you determine which major groups, if any, dominate the "bug" community. In general, a healthy stream is not dominated by only one or two major groups. In most cases, what you want to see is a diverse assemblage of "bugs."

Percent Composition of the Functional Feeding Groups

Go down the data sheet and add up the average densities of the families that represent each feeding group. Record the total average densities for each group on page 3 of the data sheet. For each group, calculate percent composition by dividing the total average density of each group by the total average density of the entire sample. (Do not use the projected density for this calculation.)

In general, all feeding groups should be well represented. No one group should dominate the sample. See the discussion on feeding groups, page 125 for more information.

EPT / Midge Ratio

This is the ratio of the number of individuals in the mayfly, stonefly, and caddisfly orders to the number of individuals in the midge family Chironomidae. Calculate the ratio by dividing the total average density of the EPT groups by the average density of the midges.

In general, the chironomid midges (also known as "true" midges), are more tolerant to pollution than EPT's. Thus the higher the EPT ratio, the better the water quality. A community with a ratio below 0.75 is considered adversely affected. (Note: this value is used by the Vermont Department of Environmental Conservation; check with your local agency biologists to determine if there is a different standard in your area.)

Family Biotic Index

Biologists have assigned **pollution tolerance values** to the different "bug" families based on their tolerance to organic pollution. The values are on a scale from 0 (least tolerant) to 10 (most tolerant.) The **biotic index** is a value of stream water quality calculated from a formula that takes into account these pollution tolerance values.

A list of pollution tolerance values for families of stream "bugs" is provided in Table 9. While this list comes from a U.S. Environmental

Protection Agency publication, the values were developed for "bug" families in the western Great Lakes and in New York State. So, if your local agency biologists can provide you with pollution tolerance values more specific to your area, use them. Go down the data sheet, multiplying the average density by the tolerance value for each family, recording the resulting values in the appropriate column. Then find the sum total of these values. Divide this sum by the total average density of the sample to calculate the family biotic index. (Do not use the projected density for this calculation). Record your result on page 3 of the data sheet.

The higher the biotic index, the more impacted the stream. R.W. Bode from the New York State Department of Environmental Conservation has developed a scale for evaluating the family biotic index. Values are interpreted as follows:

> 0-4 non impacted
> 4-6 slightly impacted
> 6-8 moderately impacted
> >8 severely impacted

Since this interpretation was developed for "bug" communities in New York State, check with your local agency biologists for assistance with adjusting these values to fit your region. Note also that this index is based on "bug" tolerances to **organic** pollution. While it may be applicable for other types of pollutants (e.g. toxics), this has not yet been thoroughly evaluated.

Analyzing taxa richness, EPT richness, projected density, percent composition of major groups and feeding groups, EPT/midge ratio, and the family biotic index will help you get a comprehensive picture of how the "bug" community responds to the condition of your stream. This procedure includes these seven different metrics because no single analysis applies to all stream conditions.

As discussed in the "Field Procedure: Macroinvertebrate Survey," you will also need to monitor other factors along with the "bugs," such as the condition of the stream bottom, the depth and velocity of the water in the stream, daylength, temperature, and water quality factors such as pH, dissolved oxygen, organic nutrients, bacteria, and toxics. As you do so, over time, you will become more familiar with the character of your stream and be better able to sort out the natural variation of your stream from human impacts.

Name_____ Group_____ Date_____ Time_____
Stream Name_____ Reach Name/#_____ Site Name_____
Site Description: Habitat type_____ Avg. depth_____
Avg. current velocity_____ Other comments_____
Site Location (lat. & long., transect #, and/or river mile_____
Driving/Hiking Directions_____
_____

Weather Conditions: ❑ Clear ❑ Cloudy ❑ Rain ❑ Other _____
Air Temp. ____ Water Temp. ____Recent weather trends:_____

Families in major groups	feeding group	density (D)			avg. D	tolerance value	tolerance X avg. D	avg. D (major group)
		rep.1	rep.2	rep.3				
mayflies							→	
stoneflies							→	
caddisflies							→	
total avg. density EPT (mayfly, stonefly, caddisfly) →								

Families in major groups	tolerance value	density (D)			avg. D	feeding group	tolerance X avg. D	avg. D (major group)
		rep.1	rep.2	rep.3				
midges (fam. chironomidae only)								
other true flies							→	
dobsonflies, alderflies, fishflies							→	
beetles							→	
dragonflies, damselflies							→	
crustaceans							→	
snails							→	
clams, mussels							→	
worms, leeches							→	
others								
total tolerance x D							→	

total average density of sample ←

Data Summary + Analysis

1. Taxa Richness (total # taxa) = ____

2. EPT Richness (total mayfly, stonefly, caddisfly taxa) = ____

3. Projected Average Density for the sample = average of the projected densities for each replicate:

replicate # (total # bugs picked/total # squares picked) x 12

1	(/) x 12 =	
2	(/) x 12 =	
3	(/) x 12 =	

average = ____

4. Percent Composition of Major Groups $\left(\dfrac{\text{total avg. density for each major group}}{\text{total avg. density of sample}}\right)$

____	mayflies	____	crustaceans
____	stoneflies	____	snails
____	caddisflies	____	clams, mussels
____	midges	____	worms, leeches
____	other true flies	____	others
____	dobsonflies, alderflies, etc.		
____	beetles		

5. Percent Composition of Functional Feeding Groups

	Shredders	Scrapers	Gathering Collectors	Filtering Collectors	Predators
total = sum avg. # density for each					
%= tot / tot. avg. density of sample					

6. EPT/Midge Ratio $\dfrac{\text{Total. Avg. density EPT}}{\text{Total Avg. density midges}}$ = ____

7. Family Biotic Index = $\dfrac{\text{the sum of all "tolerance X D" Columns}}{\text{Total average density of sample}}$ = ____

Drawings in this key are from: Merrit-Cummins: *An Introduction to the AquaticInsects of North America,* Copyright 1977 by Kendall/Hunt Publishing Company; Izaak Walton League of America (IWL); or McCafferty: *Aquatic Entomology,* © 1981 Boston: Jones and Bartlett Publishers. Reprinted with permission.

1. A. Segmented legs................................go to 2
   B. No segmented legs.........................go to 14

2. A. 6 legs..go to 3
   B. More than 6 legs..........................go to 23

3. A. No wings, or wings not fully developed and do not cover entire body........go to 4

   B. Wings cover entire body (but not legs), may appear beetle-like.................go to 26

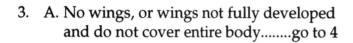

side

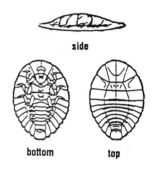

bottom   top

4. A. Body longer than it is wide............go to 5

   B. Body oval & flat; head & legs concealed beneath body.................**WATER PENNY**
   (a type of beetle larva)
   Order Coleoptera, Family Psephenidae
   Feeding Group: SCRAPER

1/4"
(excluding tails)

(all from IWL)

5. A. 2 or 3 distinct *hairlike* tails; tails not fleshy or hooked, but may be fringed with hairs...go to 6

   B. Not as above...................................go to 7

6. A. 2-3 tails; platelike or hairlike gills along sides of abdomen........**MAYFLY LARVA**
   Order Ephemeroptera
   Feeding Group: VARIES*

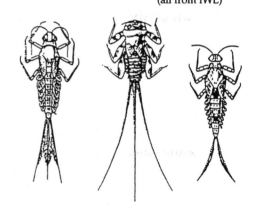

(center from IWL)   1/4"-1"
(excluding tails)

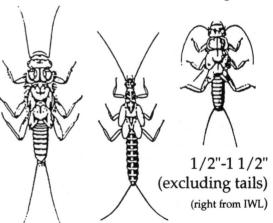

   B. 2 tails; may have hairy gills under thorax......................**STONEFLY LARVA**
   Order Plecoptera
   Feeding Group: VARIES*

1/2"-1 1/2"
(excluding tails)

(right from IWL)

* If feeding group varies, see picture key on pages 155-157 for more information.
(All drawings on this page are from McCafferty: *Aquatic Entomology,* except as noted)

7. A. 3 oar-shaped tails (gills) at *end* of abdomen; no gills along *sides* of abdomen
...............................**DAMSELFLY LARVA**
Order Odonata, Suborder Zygoptera
Feeding Group: PREDATOR

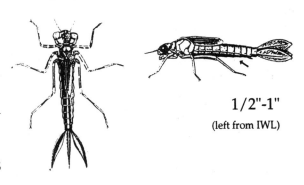

1/2"-1"
(left from IWL)

B. Not as above...................................go to 8

8. A. Fat abdomen; large eyes; mask-like lower lip............**DRAGONFLY LARVA**
Order Odonata, Suborder Anisoptera
Feeding Group: PREDATOR

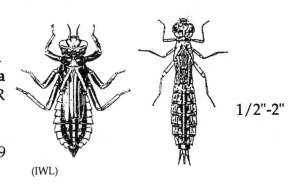

1/2"-2"

B. Not as above...................................go to 9

(IWL)

9. A. May be hiding in case made of gravel or plant parts; abdomen ends in pair of prolegs which may be hidden by hairs; each proleg has single hook on end, sometimes fused together
...............................**CADDISFLY LARVA**
Order Trichoptera
Feeding Group: VARIES*

(IWL)

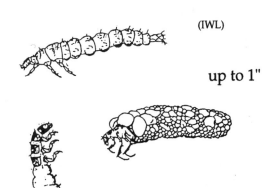

up to 1"

B. Not as above...................................go to 10

10. A. Well developed lateral filaments extend from abdominal segments..........go to 11

B. No lateral filaments along abdomen; body is hardened & stiff; tip of abdomen has small plate-like opening with hooks and filaments.
........................**RIFFLE BEETLE LARVA**
Order Coleoptera, Family Elmidae
Feeding Group: GATHERER COLLECTOR

1/4"-1/2"

*If feeding group varies, see picture key on page 158-161 for more information
(all drawings on this page from McCafferty: *Aquatic Entomology*, except as noted)

11. A. Fluffy or branched gill tufts under
abdomen..............**DOBSONFLY LARVA**
("Hellgrammite")
Order Megaloptera, Family Corydalidae
Feeding Group: PREDATOR

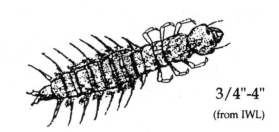

3/4"-4"
(from IWL)

B. Not as above..................................go to 12

12. A. Abdomen ends in single, unforked,
long, hairlike tail....**ALDERFLY LARVA**
Order Megaloptera, Family Sialidae
Feeding Group: PREDATOR

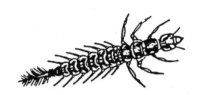

up to 1"

B. Not as above..................................go to 13

13. A. Abdomen ends in a pair of prolegs, each
with 2 hooks................**FISHFLY LARVA**
Order Megaloptera, Family Corydalidae
Feeding Group: PREDATOR

up to 1 1/2"

B. Not as above; large, obvious mouthparts
......................**AQUATIC BEETLE LARVA**
Order Coleoptera
Feeding Group: PREDATOR

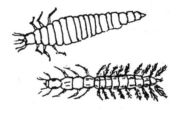

1/4"-1"
(from IWL)

14. A. Has small but distinct head; body less
than 1/2" long................................go to 15

B. Appears not to have a head, although it
may be retracted into body..........go to 16

15. A. Body widens at bottom end (bowling
pin shaped); may be attached to substrate; dark head....**BLACK FLY LARVA**
Order Diptera, Family Simuliidae
Feeding Group: FILTERER COLLECTOR

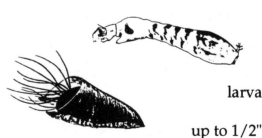

larva

up to 1/2"

pupa

(all drawings on this page from McCafferty: *Aquatic Entomology*, unless otherwise noted)

The Streamkeeper's Field Guide

15. B. Both ends of body about the same
width; tiny pair of prolegs under head
& at tip of abdomen......**MIDGE LARVA**
    Order Diptera, Family Chironomidae
    Feeding Group: GATHERER COLLECTOR

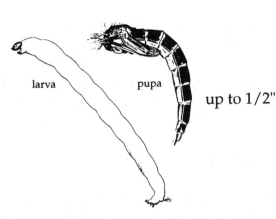

larva   pupa

up to 1/2"

16. A. Fleshy Caterpillar-like body........go to 17
    B. Body not caterpillar-like...............go to 18

17. A. Two feathered "horns" at back end;
caterpillar-like legs
...................**WATERSNIPE FLY LARVA**
    Order Diptera, Family Athericidae
        Feeding Group: PREDATOR

1/4"-1"

last segment not
swollen -
SHREDDER

B. Can be up to 4" long; head not apparent
because it is retracted into body; may have
fleshy, finger-like extensions at one end
................................**CRANEFLY LARVA**
    Order Diptera, Family Tipulidae
Feeding Group: SHREDDER OR PREDATOR

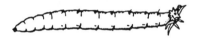

1/3"-4"

last segment swollen or
no fingerlike extensions
- PREDATOR
(all from Cummins)

18. A. Body without hard shell............ ..go to 19
    B. Body with hard shell.....................go to 21

19. A. Flattened, unsegmented, worm-like
body; distinct eye spots; gliding movement..................................**PLANARIAN**
                (Flatworm)
            Class Turbellaria
    Feeding Group: PREDATOR or PARASITE

up to 3/4"
(from IWL)

B. Segmented body..........................go to 20

20. A. Flattened body with suckers at each end
....**LEECH**
            Class Hirudinea
    Feeding Group: PREDATOR or PARASITE

1/4"-2"
(bottom figure
from IWL)

(all drawings on this page are from McCafferty: *Aquatic Entomology,* **unless otherwise noted)**

20. B. Segmented, earthworm-like body
......................**AQUATIC EARTHWORM**
Class Oligochaeta
Feeding Group: GATHERER COLLECTOR

1/4"-2"

21. A. Snail-like...go to 21

B. Body enclosed within two hinged shells
.......**FRESHWATER CLAM or MUSSEL**
Class Pelecypoda
Feeding Group: FILTERER COLLECTOR

(both from IWL)

22. A. Has operculum (hard covering used to
close the opening)..........**GILLED SNAIL**
Class Gastropoda, Order Prosobranchia
Feeding Group: SCRAPER

(right from IWL)

B. No operculum; may be spiral-shaped,
limpet-like, or coiled in one plane
...................**LUNG-BREATHING SNAIL**
Class Gastropoda, Order Pulmonata
Feeding Group: SCRAPER

(both from IWL)

23. A. Looks like spider; may be very tiny; has
8 legs.............................**AQUATIC MITE**
Class Arachnida, Order Hydracarina
Feeding Group: PREDATOR

up to 1/8"

B. Not as above...................................go to 24

24. A. Lobster or shrimp-like..................go to 25

B. Armadillo shaped body, wider than
high; crawls slowly on bottom
...............................**AQUATIC SOWBUG**
Subphylum Crustacea, Order Isopoda
Feeding Group: SHREDDER

1/4"-3/4"

(all drawings on this page are from McCafferty: *Aquatic Entomology*, unless otherwise noted)

25. A. Looks like tiny shrimp; swims quickly
on its side.................................SCUD
Subphylum Crustacea, Order Amphipoda
Feeding Group: SHREDDER

1/4"-1/2"

B. Looks like small lobster; has 2 large
front claws (10 legs total)......**CRAYFISH**
Subphlum Crustacea, Order Decapoda
Feeding Group: GATHERER COLLECTOR

up to 6"

26. A. Beetle-like, crawls slowly on bottom
.............................**RIFFLE BEETLE ADULT**
Order Coleoptera, Family Elmidae
Feeding Group.: SCRAPER or
GATHERER COLLECTOR

1/4"

B. Beetle-like, swims quickly............go to 27

27. A. Wings meet along the midline of back
side of body, they do not overlap
....**BEETLE ADULT**
Order Coleoptera
Feeding Group: MOST ARE PREDATORS

up
to
1"

Predaceous
Diving Beetle

Whirligig Beetle
(swims in circular motion on
water surface)

27. B. Wings overlap on backside, usually
form a visible triangular pattern just
below head.....................................go to 28

28. A. Front legs are shorter than mid and hind
legs; propels itself with oar-like strokes,
................................**WATER BOATMAN**
Order Hemiptera, Family Corixidae
Feeding Group: VARIES

up to 3/4"

B. Similar to backswimmer but swims
upside down, on its back
.....................................**BACKSWIMMER**
Order Hemiptera, Family Notonectidae

Feeding Group: PREDATOR

up to 3/4"

(all drawings on this page are from McCafferty: *Aquatic Entomology*, unless otherwise noted)

A PICTURE KEY TO MAYFLY, STONEFLY AND CADDISFLY FAMILIES

Drawings in this key are from: Merrit-Cummins: *An Introduction to the Aquatic Insects of North America*. Copyright 1977 by Kendall/Hunt Publishing Company; Izaak Walton League of America (IWL); or McCafferty: *Aquatic Entomology*, © 1981 Boston: Jones and Bartlett Publishers. Reprinted with permission.

The common names used in this key e.g.,"flathead mayfly," "green stonefly," or "common netspinner" are taken from McCafferty: Aquatic Entomology,© 1981 Boston: Jones and Bartlett Publishers. Be aware that common names for aquatic "bugs" may vary from region to region, and even from person to person. The scientific names, which do not vary, appear next to the common names.

Mayflies (Order Ephemeroptera)

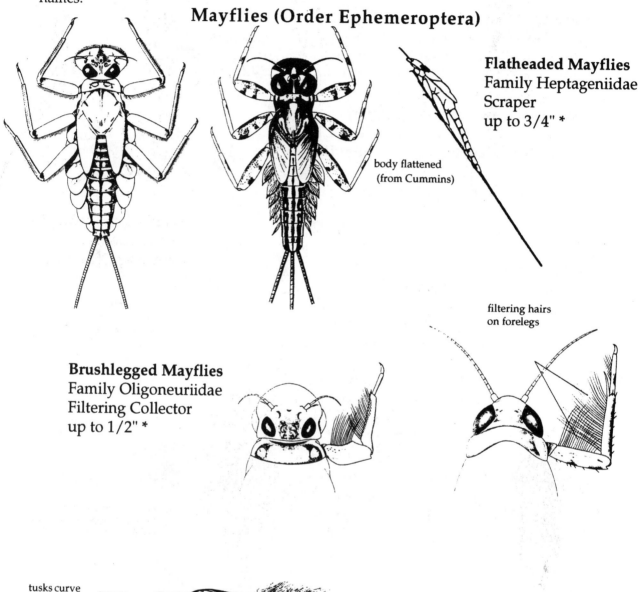

Flatheaded Mayflies
Family Heptageniidae
Scraper
up to 3/4" *

body flattened
(from Cummins)

filtering hairs
on forelegs

Brushlegged Mayflies
Family Oligoneuriidae
Filtering Collector
up to 1/2" *

tusks curve
up and out

Common Burrowers
Family Ephemeridae
Gathering Collector
1/2"-1 1/4" *

*sizes do not include tails
(All drawings on this page are from McCafferty: *Aquatic Entomology*, unless otherwise noted)

Mayflies (Order Ephemeroptera)

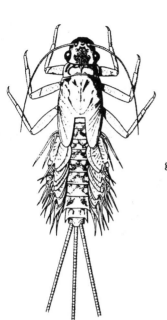

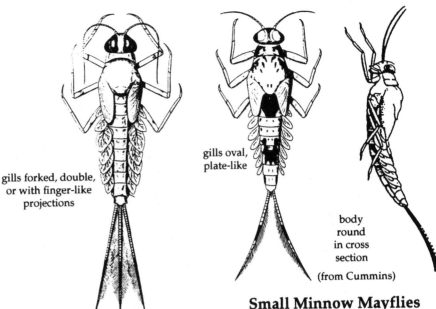

gills forked, double, or with finger-like projections

gills oval, plate-like

body round in cross section

(from Cummins)

Small Minnow Mayflies
Family Baetidae
Gathering Collector
up to 1/2" *

Pronggills
Family Leptophlebiidae
Gathering Collector
up to 5/8" *

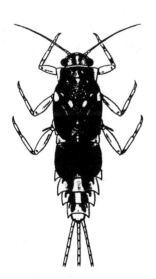

gills on abdominal segment 2 overlapping

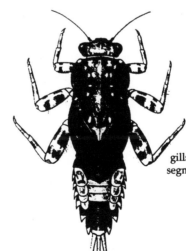

gills begin on abdominal segment 3; gills lacking on segments 1 & 2

Small Squaregills
Family Caenidae
Gathering Collector
up to 3/8" *

Spiny Crawlers
Family Ephemerellidae
Gathering Collector
up to 5/8" *

*sizes do not include tails

(all drawings on this page are from McCafferty: *Aquatic Entomology,* unless otherwise noted)

A PICTURE KEY TO MAYFLY, STONEFLY, AND CADDISFLY FAMILIES

Stoneflies (Order Plecoptera)

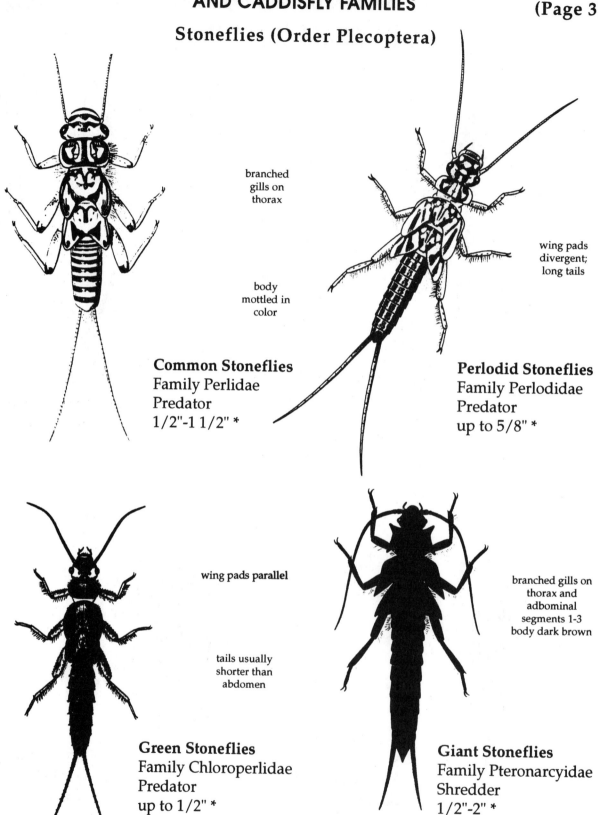

branched gills on thorax

body mottled in color

wing pads divergent; long tails

Common Stoneflies
Family Perlidae
Predator
1/2"-1 1/2" *

Perlodid Stoneflies
Family Perlodidae
Predator
up to 5/8" *

wing pads parallel

tails usually shorter than abdomen

branched gills on thorax and adbominal segments 1-3 body dark brown

Green Stoneflies
Family Chloroperlidae
Predator
up to 1/2" *

Giant Stoneflies
Family Pteronarcyidae
Shredder
1/2"-2" *

(all drawings on this page are from McCafferty: *Aquatic Entomology,* **unless otherwise noted)**

The Streamkeeper's Field Guide

Stoneflies (Order Plecoptera)

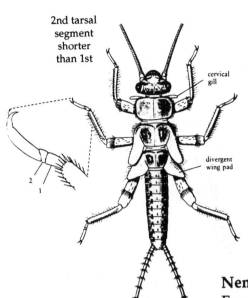

2nd tarsal segment shorter than 1st

cervical gill

divergent wing pad

Nemourid Broadbacks
Family Nemouridae
Shredder
up to 1/2" *

1st and 2nd tarsal segments about the same length

divergent wing pad

Taeniopterygid Broadbacks
Family Taeniopterygidae
Shredder
up to 1/2" *

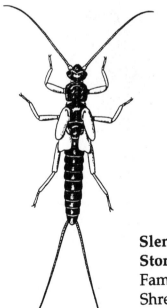

Slender Winter Stoneflies
Family Capniidae
Shredder
up to 1/4" *

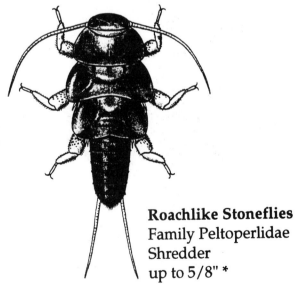

groove on underside of abdomen

Roachlike Stoneflies
Family Peltoperlidae
Shredder
up to 5/8" *

*sizes do not include tails
(all drawings from McCafferty: *Aquatic Entomology*, unless otherwise noted)

Caddisflies (Order Trichoptera)

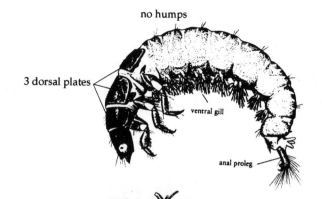

no humps

3 dorsal plates

ventral gill

anal proleg

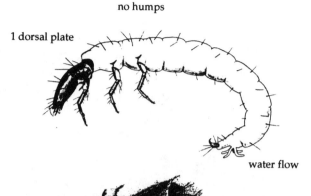

no humps

1 dorsal plate

water flow

Common Netspinners
Family Hydropsychidae
Filtering Collector
up to 1 1/4"

(Merrit & Cummins)

Fingernet Caddisflies
Family Philopotamidae
Filtering Collector
up to 1/2"

plate-like appendage just below head

thorn-like appendage just below head

(Merrit & Cummins)

Trumpetnet Caddisflies

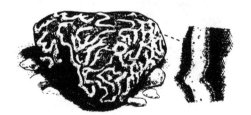

(Merrit & Cummins)

Nettube Caddisflies
Family Psychomyiidae
Filtering Collector
up to 1/2"

Tubemaking Caddisflies
Both: Family Polycentropodidae
Filtering Collector
up to 1"

(Drawings on this page are from McCafferty: *Aquatic Entomology,* unless otherwise noted)

Caddisflies (Order Trichoptera)

1 dorsal plate

no humps

Freeliving Caddisflies
Family Rhyacophilidae
Predator
up to 1"

to distinguish freeliving caddisflies from netspinners that have been dislodged from their nets: freeliving caddisflies have a long, small head that is narrower than the thorax; their body is often bright green; the head of netspinners is as wide as the thorax

has conspicuous antenna

2 dorsal plates, no humps

Humpless Case Makers
Family Brachycentridae
Filtering Collector
up to 1/2"

Longhorned Case Makers
Family Leptoceridae
Gathering Collector
up to 5/8"

(Drawings on this page are from McCafferty: *Aquatic Entomology*, unless otherwise noted)

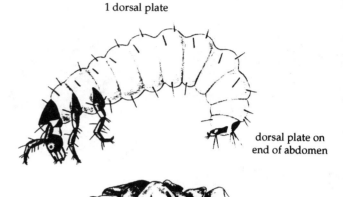

1 dorsal plate

dorsal plate on end of abdomen

dorsal plate

anal prolegs

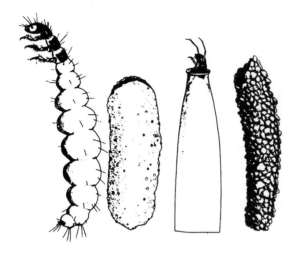

Micro Caddisflies
Family Hydroptilidae
Scraper
up to 1/4"

Saddlecase Makers
Family Glossosomatidae
Scraper
up to 1/2"

body very rounded

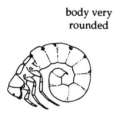

dorsal hump

lateral hump

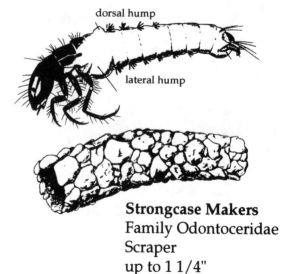

Snailcase Makers
Family Helicopsychidae
Scraper
up to 1/2"

Strongcase Makers
Family Odontoceridae
Scraper
up to 1 1/4"

(all drawings on this page are from McCafferty: *Aquatic Entomology,* unless otherwise noted)

Caddisflies (Order Trichoptera)

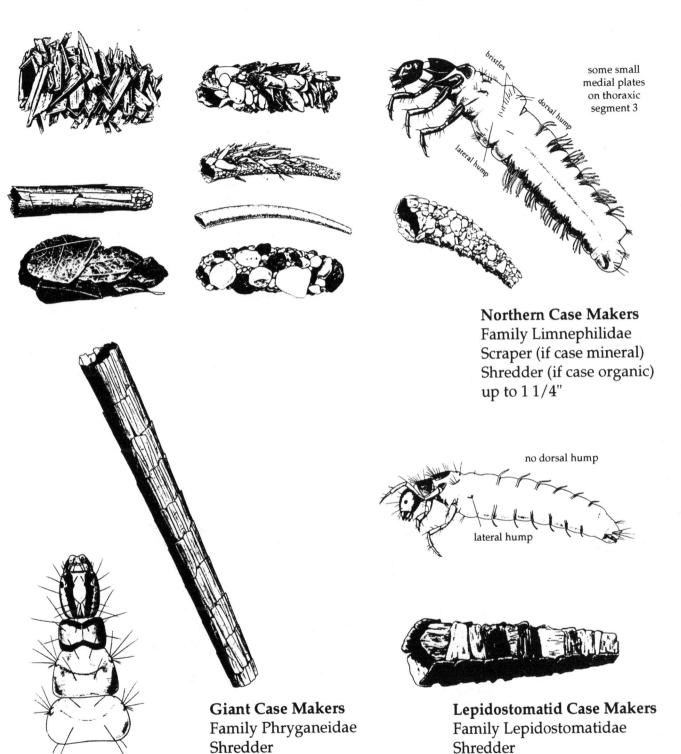

bristles

some small medial plates on thoraxic segment 3

dorsal hump

lateral hump

Northern Case Makers
Family Limnephilidae
Scraper (if case mineral)
Shredder (if case organic)
up to 1 1/4"

no dorsal hump

lateral hump

Giant Case Makers
Family Phryganeidae
Shredder
up to 1 1/2"

Lepidostomatid Case Makers
Family Lepidostomatidae
Shredder
up to 1/2"

(all drawings on this page are from McCafferty: *Aquatic Entomology*, unless otherwise noted)

POLLUTION TOLERANCE VALUES FOR FAMILIES OF STREAM MACROINVERTEBRATES

Values are based on a family's tolerance to organic pollution. The scale ranges from 0 (least tolerant) to 10 (most tolerant). This chart is adapted from U.S. Environmental Protection Agency, *Rapid Bioassessment Protocols for Use in Streams and Rivers, Benthic Macroinvertebrates and Fish*, May 1989. Values were developed for "bug" families in the Great Lakes and New York State. Before using this chart, check with your local biologists to determine if values have been developed for your specific region.

Order	Common Name	Family	Tolerance Value
Stoneflies (Plecoptera)	Common Stoneflies	(Perlidae)	1
	Green Stoneflies	(Chloroperlidae)	1
	Giant Stoneflies	(Pteronarcyidae)	0
	Nemourid Broadbacks	(Nemouridae)	2
	Perlodid Stoneflies	(Perlodidae)	2
	Rolledwinged Stoneflies	(Leuctridae)	0
	Slender Winter Stoneflies	(Capniidae)	1
	Taeniopterygid Broadbacks	(Taeniopterygidae)	2
Mayflies (Ephemeroptera)	Armored Mayflies	(Baetiscidae)	3
	Brushlegged Mayflies	(Oligoneuriidae)	2
	Cleftfooted Minnow Mayflies	(Metretopodidae)	2
	Common Burrowers	(Ephemeridae)	4
	Flatheaded Mayflies	(Heptageniidae)	4
	Hacklegills	(Potamanthidae)	4
	Little Stout Crawlers	(Tricorythidae)	4
	Pale Burrowers	(Polymitarcyidae)	2
	Primitive Minnow Mayflies	(Siphlonuridae)	7
	Pronggills	(Leptophlebiidae)	2
	Small Minnow Mayflies	(Baetidae)	4
	Small Squaregills	(Caenidae)	7
	Spiny Crawlers	(Ephemerellidae)	1
Caddisflies (Trichoptera)	Bushtailed Case Makers	(Sericostomatidae)	3
	Common Netspinners	(Hydropsychidae)	4
	Fingernet Caddisflies	(Philopotamidae)	3
	Freeliving Caddisflies	(Rhyacophilidae)	0
	Giant Case Makers	(Phryganeidae)	4
	Hoodcase Makers	(Molannidae)	6
	Humpless Case Makers	(Brachycentridae)	1
	Lepidostomatid Case Makers	(Lepidostomatidae)	1
	Longhorned Case Makers	(Leptoceridae)	4
	Micro Caddisflies	(Hydroptilidae)	4
	Nettube Caddisflies	(Psychomyiidae)	2
	Northern Case Makers	(Limnephilidae)	4
	Saddlecase Makers	(Glossosomatidae)	0
	Snailcase Makers	(Helicopsychidae)	3
	Strongcase Makers	(Odontoceridae)	0
	Trumpetnet & Tubemaking Caddisflies	(Polycentropodidae)	6
Dobsonflies, Alderflies, & Fishflies (Megaloptera)	Dobsonflies & Fishflies	(Corydalidae)	0
	Alderflies	(Sialidae)	4

POLLUTION TOLERANCE VALUES FOR FAMILIES OF STREAM MACROINVERTEBRATES

Order (or Class)	Common Name	Tolerance Value
True Flies (Order Diptera)	Aquatic Dance Flies.....................	(Empididae)........ 6
	Aquatic Longlegged Flies..........	(Dolichopodidae).........4
	Aquatic Muscids.........................	(Muscidae)...................6
	Biting Midges.............................	(Ceratopogonidae).......6
	Black Flies..................................	(Simuliidae)..................6
	Craneflies...................................	(Tipulidae)...................3
	Horse & Deer Flies......................	(Tabanidae)..................6
	Moth Flies..................................	(Psychodidae)............10
	Netwinged Midges......................	(Blephariceridae)..........0
	Rattailed Maggots.......................	(Syrphidae)................10
	Shore Flies.................................	(Ephydridea)................6
	True Midges (blood red).............	(Chironomidae)............8
	True Midges (others, including pink)	(Chironomidae)............6
	Watersnipe Flies.........................	(Anthericidae)..............2
Beetles (Order Coleoptera)	Water Pennies.............................	(Psephenidae)...............4
	Longtoed Water Beetles.............	(Dryopidae)..................5
	Riffle Beetles...............................	(Elmidae).....................4
Damselflies & Dragonflies (Order Odonata)	Broadwinged Damselflies..........	(Calopterygidae)............5
	Narrowwinged Damselflies........	(Coenagrionidae)..........9
	Spreadwinged Damselflies..........	(Lestidae).....................9
	Belted & River Skimmers.............	(Macromiidae)..............3
	Biddies......................................	(Cordulegastridae)......3
	Clubtails....................................	(Gomphidae).................1
	Common Skimmers......................	(Libellulidae)................9
	Darners.....................................	(Aeshnidae)..................3
	Greeneyed Skimmers...................	(Corduliidae)................5
Crustaceans (Class Crustacea)	Aquatic Sowbugs (Order Isopoda)	8
	Crayfish (Order Decapoda)...........	6
	Scuds (Order Amphipoda)...........	(Gammaridae)..............4
		(Talitridae)...................8

Snails (Class Gastropoda) 6-8*

Aquatic Earthworms (Class Oligochaeta)..6-10*

Leeches (Class Hirudinea)...10

Planarians (Class Turbellaria)...4

* Depending on the genus, tolerance values for snails vary from 6-8. For aquatic worms, they vary from 6-10. See U.S. EPA publication and local aquatic entomologists for more information.

Note: The common names used in this chart, e.g., "flatheaded mayfly," "green stonefly," or "common netspinner" are taken from McCafferty: *Aquatic Entomology*, © 1981 Boston: Jones and Bartlett Publishers. Reprinted by permission. Be aware that common names for aquatic "bugs" may vary from region to region, and even from person to person. The scientific names, which do not vary, appear next to the common names.

Water Quality

At this point, you have inventoried your watershed, mapped and surveyed your stream reach, evaluated the physical nature of your stream, and become intimate with the spineless inhabitants of the bottom. This chapter introduces you to a final aspect of your stream survey: chemical water quality. While understanding the physical and biological life of a stream is vital to evaluating its health, chemical water quality is also an important piece of the puzzle.

Evaluating the chemical water quality of a stream involves looking at the concentration of dissolved and suspended substances in the water. Some of these substances come from the atmo-sphere, delivered to streams via precipitation. Other substances are picked up from soil, vegetation and other sources, and carried to streams via surface runoff. Groundwater also picks up chemical constituents from contact with underlying rocks and sediment, and eventually delivers them to streams as subsurface runoff.

The concentration of substances in the water depends on many factors, including natural and human introduced phenomena. Concentrations vary naturally from stream to stream, and from site to site on the same stream, depending on the geology, climate, soil and vegetation of the watershed. The concentrations vary throughout the year, from season to season, from day to day, and sometimes from hour to hour.

Many substances increase or decrease with the timing and quantity of runoff. During dry weather, streams receive much of their flow from groundwater. Thus, the concentration of minerals and salts may be greater in dry times than during wet weather, when increased surface runoff and stream flow may dilute the concentration of substances. However, if there are erosion and non-point pollution sources in a watershed, increased surface runoff will also carry increased concentrations of sediment and pollutants into streams.

The chemical water quality of a stream is good if naturally occurring substances are present in the concentrations appropriate for the particular stream ecosystem and the life it supports. Problems occur when human activities alter the concentration of naturally occurring substances or introduce foreign substances that may be toxic to stream life.

This chapter focuses on the three most basic water quality parameters: **pH, dissolved oxygen,** and **temperature.** pH is a measure of how acidic or basic the water is. Dissolved oxygen is the amount of oxygen dissolved in the water, and thus available for aquatic organisms to use. Although not a chemical parameter, temperature is always included in water quality monitoring because it affects the concentrations and reactivity of many other parameters.

In addition to being the most basic water quality parameters, pH, dissolved oxygen, and temperature are also easy to measure and require relatively inexpensive equipment. We recommend you start your monitoring with these three parameters. As you delve more deeply into your watershed inventory and stream monitoring, you may add other parameters to your repertoire. Your strategy will depend on the nature of your particular stream and the human impacts it faces.

Other commonly monitored chemical water quality parameters include alkalinity, conductivity, nutrients such as nitrogen and phosphorus, metals, detergents, and petroleum hydrocarbons. Parameters that are not chemi-cal in nature but are also commonly monitored include bacteria, biological oxygen demand (BOD), and turbidity.

Some of the parameters listed above require more expensive equipment and/or laboratory analysis. However, many are included in volunteer monitoring programs. They are defined and discussed in the last part of this chapter.

Miss Mayfly Revisited

As always, consult Miss Mayfly's Guide on page 34 before heading out into the field. When you monitor water quality parameters you will be using various field test kits and reagents. So, here are a few additional safety items for you to consider:

- Make sure you read the material handling and safety information that is provided with each test kit you purchase.
- Many reagents are irritable to the eyes. We recommend you wear protective eyewear such as laboratory glasses or goggles when conducting tests.
- Some reagents are irritable to the skin. We recommend you wear disposable latex surgical gloves when conducting tests.
- Always have a jug of clean water and a wash bucket on hand to use for rinsing skin or eyes in case of contact with chemical reagents.
- Wash your hands thoroughly after conducting the tests.

pH (PARTS HYDROGEN)

Acids and bases are defined by the activity of two very reactive ions: hydrogen ions (H+) and hydroxyl ions (OH-). A solution that has more hydrogen ion activity than hydroxyl activity is considered acidic; one that has more hydroxyl ion activity than hydrogen ion activity is considered basic.

pH is an important limiting chemical factor for aquatic life. If the water in a stream is too acidic or too basic, the H+ or OH- ion activity may disrupt crucial biochemical reactions, harming or killing stream organisms.

pH is expressed in a scale which ranges from 1 to 14. A solution with a pH value less than 7 has more H+ activity than OH-, and thus is considered acidic. A solution with a pH value greater than 7 has more OH- activity than H+, and thus is considered basic. A solution with a pH of 7 is considered neutral; the H+ and OH- activity is balanced.

The pH scale is logarithmic. This means that as you go up and down the scale, the values change in factors of ten. A one-point pH change indicates the strength of the acid or base has increased or decreased tenfold; a two-point pH change indicates a 100 fold change in acidity or basicity; a 3 point pH change indicates a 1000 fold change in acidity or basicity, and so on.

For example, lye is a very strong base (pH ~ 13.5). It is 10 times more basic than bleach (pH ~ 12.5), 100 times more basic than ammonia (pH ~ 11.5), and 100,000 times more basic than baking soda (pH ~ 8.5).

Pure distilled water, at 25^{0}C, has a neutral pH value of 7. Natural, unpolluted rainwater tends to be as acidic as 5.6 pH, because it absorbs carbon dioxide as it falls through the atmosphere, forming carbonic acid. Streams generally have pH values ranging between 6 and 9, depending on the presence of dissolved substances that come from bedrock, soils, and other materials in the watershed.

Where soils are **alkaline** (basic), stream pH levels may be greater than 7. **Buffering** is the ability of a solution to receive large amounts of an acid or base and not undergo an appreciable change in pH. Often the alkaline soil and mineral particles washed into streams buffer the affect of rainwater acidity.

Each individual stream tends to have a narrow range of pH values. Many stream organisms adapt to the specific pH range of their home waters. For some critters, a small change in pH can be lethal, while others can tolerate a broader range. As figure 7.1 indicates, salmonids are among the most sensitive organisms to changes in pH.

Human activities influence the pH of our surface water bodies. Air pollution from car exhaust and other fossil fuel burning has increased the concentrations of sulfur and nitrogen oxide in the atmosphere. These pollutants can move far from their original release sites. They become part of the hydrologic cycle and fall back to earth, with rain, as weak sulfuric acid and nitric acid.

This phenomenon is known as **acid rain**. In some areas of the U.S. and Canada, acid rain has lowered the pH in lakes below 5. This pH can be lethal for fish and other aquatic organisms. Though these lakes are very "clean looking" to the casual observer, they are devoid of most life.

Changes in pH also affect aquatic organisms indirectly by changing other aspects of water chemistry. For example, as pH increases (acidity decreases), smaller amounts of ammonia are needed to reach a level that is toxic to fish. As pH decreases (acidity increases), the concentration of metals may increase because higher acidity increases their ability to be dissolved from sediments into the water. Metals such as copper and aluminum can disrupt the function of fish gills or cause developmental deformities.

TABLE 9
pH OF COMMON SUBSTANCES AND
LETHAL pH LIMITS FOR AQUATIC ORGANISMS

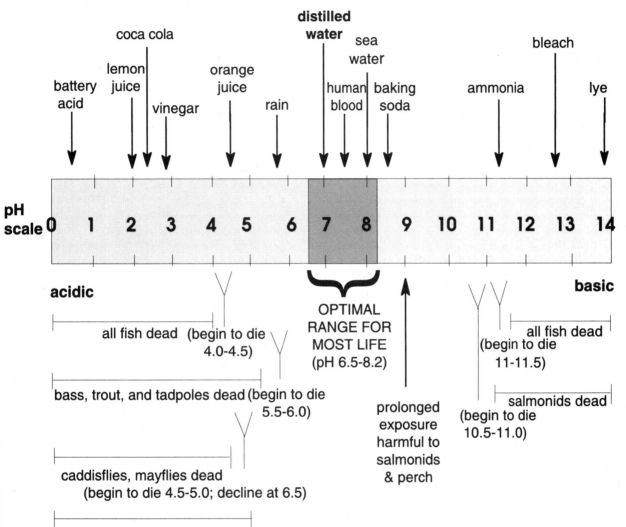

Compiled from Campbell, G, and Wildberger, S. 1992. *The Monitor's Handbook.* LaMotte Company, Chertertown, MD and MacDonald, L.H. with Smart, A.W. and Wissman, R.C. 1991. *Monitoring Guidelines to Evaluate Effects of Forestry Activity on Streams in the Pacific Northwest and Alaska.* U.S. Environmental Protection Agency, Region 10, Seattle, WA.

FIELD PROCEDURE: PH

EQUIPMENT NEEDED

pH test kit
buffer solutions of known pH value
latex gloves
protective eye wear
distilled water
wash bottles for rinsing glassware in the field
extra jug of water for first aid in case of spill or eye contact
watertight containers for liquid wastes
container for solid wastes
towels for drying hands and equipment
Data Sheet 7, "Water Quality"
clipboard
pencils

WHAT?

Measuring pH indicates how acidic or basic the water in your stream is.

WHERE?

For baseline studies, select a site where the water sample is typical of your stream reach and does not represent a localized condition. For example, do not sample next to a discharge or drainage pipe. Try to choose a site free of suspended sediments, and deep enough so that you can collect your sample from beneath the surface of the water. The site you select for pH should be the same site you sample for other water quality parameters, as well as for macroinvertebrates, so you can compare the results of all tests with one another.

If you are trying to assess the impact of a particular discharge or other activity, you will need to consider localized conditions. Select a site just upstream from the suspected source of impact, a second site just downstream, and additional sites further downstream from the suspected source. Space out these additional downstream sites so you can determine over what distance your stream is able to recover from the impact.

Equipment for Measuring pH

Field Test Kits

pH field test kits use liquid indicator dyes that change color according to the pH of the solution. This color is then matched to color standards representing known pH values. Most groups start out with a **wide range** kit that covers pH values from about 4 to 10, in increments of 1 or 0.5.

As you become more familiar with your stream's particular pH range, you may want to get a **narrow range** test kit. Narrow range kits cover a smaller range of pH values with greater **sensitivity**. This means the overall range of values covered is less, but the kit can distinguish smaller increments of pH.

For example, if you find that your stream's pH tends to be above 8, you can use a kit with a range of 8.0 - 9.4; if your stream's pH tends to be below 7.5, you can use a kit with a range of 6.0 - 7.4; etc. Narrow range kits use different types of indicator dyes depending on the specific range. They usually detect pH changes in increments of 0.1 or 0.2.

Electronic Meters

Electronic meters can provide more precise readings over a wider range of pH values than field test kits. Their drawback is that they are much more expensive. Pocket pH meters, which are much less expensive than regular meters, are also available, but they have a tendency to wear out faster.

The methods in this section are written for field test kits because we assume most Streamkeeper groups are not able to afford electronic meters. If you conduct quality control checks on your field test kits as described in Chapter 8 your data will be reliable and useful.

If you do use a meter, make sure you **calibrate** it frequently. Calibration involves testing and adjusting your meter using standard solutions of known pH value. You should receive instructions for calibrating your meter when you purchase it.

WHEN?

For baseline information, we recommend that you measure pH at least monthly. For consistency, conduct your pH tests at the same time of the day each month. For impact assessments, you may need to monitor pH more frequently, depending on how often you suspect the source of pollution impacts your stream.

HOW?

Before you head out to the field:

1. Read the instructions that accompany your pH test kit, including safety and material handling considerations.

2. For a quality control check, try out your test kit with buffer solutions of known pH value (available from the company that sold you your kit). Make sure that all Streamkeepers are interpreting the colors the same way.

3. Clean all containers/tubes thoroughly and rinse with distilled water.

In the field:

1. Follow the directions that accompany your pH test kit.

2. Use gloves while collecting the sample and conducting the test so you don't contaminate your water sample and to protect your hands. Wear glasses or goggles to protect your eyes.

3. Rinse testing containers/tubes a few times with the water you will sample.

4. Collect samples in the main current of the stream, away from the banks, and under the surface of the water, where homogeneous mixing of the water occurs.

5. Analyze the sample immediately; temperature changes and contact with the air can affect the pH value.

6. We recommend you conduct three replicate tests at each sampling site. Record all values on Data Sheet 7, "Water Quality."

7. Be certain to carry out all waste products when you leave your stream site. The test products contain chemicals that could pollute your stream if left behind or poured back into the water. Refer to the material handling information provided with your test kits to find out about proper disposal methods.

NOW WHAT?

To report one result, average your values, discarding any **outlyers** (values that are way out of range and probably due to procedural error).

Look at the lethal pH limits for aquatic life on figure 7-1. Are your results consistent with your watershed inventory information and stream survey data? Is the pH of the stream preventing any native fish populations from thriving in your stream?

If you haven't done so already, find out what your state's pH standard is for your stream. Your stream's pH standard will depend on its classification, which is based on its "beneficial uses." Compare your results with the standard.

Research the soil types in your stream's watershed to get an understanding of what minerals may be dissolved in the water, and the expected level of natural acidity or basicity.

Alkalinity is a good follow-up test to pH, especially if your stream tends to be acidic. Alkalinity is a measure of how much a body of water is able to neutralize acids. (See section on alkalinity on page 179).

For a quality control check, **calibrate** your test kit with a pH meter. Calibration refers to checking your equipment against a known outcome to determine how accurate your results are. Plan to meet with an agency scientist or water quality technician in the field so you can test samples of water taken from the exact same locations where meter readings are taken. See page 192 in Chapter 8, "Credible Data," for more information on accuracy and calibration procedures.

DISSOLVED OXYGEN

Almost all plants and animals, whether living on land or in the water, need oxygen for their growth and survival. This life-giving gas is present in the water in a dissolved form. Compared to the atmosphere, there is a lot less oxygen available in water. Thus aquatic organisms have devised specialized means of extracting and storing oxygen from the water. Many aquatic plants have spongy tissue that enables them to store oxygen. Most aquatic animals possess gills or other types of specialized breathing adaptations.

Oxygen enters water from the air at the surface of the stream. Oxygen also enters the water from aquatic plants and algae. It is a by-product of **photosynthesis**, the process by which green plants use sunlight and carbon dioxide to produce their energy source, carbohydrates.

The amount of oxygen dissolved in water is expressed as a **concentration**. A concentration is the amount in weight (mass) of a particular substance per a given volume of liquid. The dissolved oxygen concentration in a stream is the mass of the oxygen gas present, in milligrams (mg), per liter (l) of water. Milligrams per liter (mg/l) can also be expressed as parts per million (ppm).

Natural Factors Affecting Dissolved Oxygen

The concentration of dissolved oxygen in a stream is affected by many factors:

- **Temperature:** Oxygen is more easily dissolved in cold water. Thus stream organisms that require high levels of dissolved oxygen, such as salmonids and many types of mayflies, stoneflies, and caddisflies, usually inhabit cold water streams.

- **Flow:** Oxygen concentrations vary with the volume and velocity of water flowing in a stream. Faster flowing white water areas tend to be more oxygen rich because more oxygen enters the water from the atmosphere in those areas than in slower, stagnant areas.

Also, slower moving water can heat up more from the sun, thus reducing oxygen levels.

- **Aquatic Plants:** The presence of aquatic plants in a stream affects the dissolved oxygen concentration. As mentioned above, green plants release oxygen into the water during photosynthesis. Photosynthesis occurs during the day when the sun is out and ceases at night. Thus in streams with significant populations of algae and other aquatic plants, the dissolved oxygen concentration may fluctuate daily, reaching its highest levels in the late afternoon. Because plants, like animals, also take in oxygen, dissolved oxygen levels may drop significantly by early morning.

- **Altitude:** Oxygen is more easily dissolved into water at low altitudes than at high altitudes.

- **Dissolved or suspended solids:** Oxygen is also more easily dissolved into water with low levels of dissolved or suspended solids. Thus salt water tends to have lower concentrations of dissolved oxygen than fresh water.

Human Activities Affecting Dissolved Oxygen

There are many ways our activities influence the amount of oxygen dissolved in stream water:

- **Removal of riparian vegetation** may lower oxygen concentrations due to increased water temperatures resulting from a lack of canopy shade and increased suspended solids resulting from erosion of bare soil.

- **Typical urban human activities** may lower oxygen concentrations. Runoff from impervious surfaces bearing salts, sediments, and other pollutants increases the amount of suspended and dissolved solids in stream water.

- **Organic wastes and other nutrient inputs** from sewage and industrial discharges, septic tanks, and agricultural and urban runoff can

The Streamkeeper's Field Guide

result in decreased oxygen levels. Nutrient inputs often lead to excessive algal growth. When the algae die, the organic matter is decomposed by bacteria. Bacterial decomposition consumes a great deal of oxygen.

• **Dams** may pose an oxygen supply problem when they release waters from the bottom of their reservoirs into streams and rivers. Although the water on the bottom is cooler than the warm water on top, it may be low in oxygen if large amounts of organic matter (in the form of plant and animal remains) has fallen to the bottom and has been decomposed by bacteria.

Effects on Aquatic Life

If the dissolved oxygen concentration falls too low due to any of the above factors, a stream may not be able to support aquatic life. Most stream organisms have specific oxygen requirements and can only live in streams or in areas of streams that meet their needs. Some organisms, such as salmonids, and many species of mayflies, stone-flies, and caddisflies, require high levels of oxygen, while others can get by with much less. Each organism prefers to live in the conditions to which it is best adapted.

Usually streams with high dissolved oxygen concentrations (greater than 8 mg/l) are considered healthy systems. They are able to support a greater diversity of aquatic organisms. They are typified by cold, clear water, with enough riffles to provide sufficient mixing of atmospheric oxygen into the water.

In streams that have been impacted by any of the factors discussed above, summer is usually the most crucial time for dissolved oxygen levels because stream flows tend to lessen and water temperatures tend to increase. For example, when dissolved oxygen levels fall below 6 mg/l in trout, a fish kill can result.

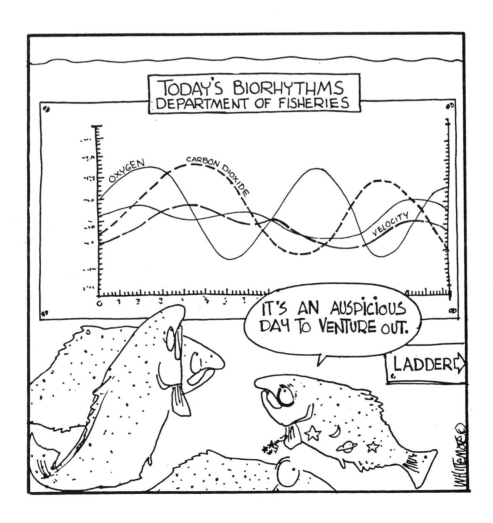

FIELD PROCEDURE: DISSOLVED OXYGEN

EQUIPMENT NEEDED

dissolved oxygen test kit (see watershed window)
latex gloves
protective eye wear
distilled water
wash bottles for rinsing glassware in the field
extra jug of water for first aid in case of spill or eye contact
watertight containers for liquid wastes
container for solid wastes
towels for drying hands and equipment
Data Sheet 7, "Water Quality"
clipboard
pencils

WHAT?

In this procedure, you will measure the amount of oxygen in the stream available for aquatic life.

Equipment for Measuring Dissolved Oxygen

Field Test Kits

Field test kits use a modification of the Winkler Method, the standard method for measuring dissolved oxygen. The method involves a titration. A solution of known strength, the **titrant**, is added to a specific volume of a treated sample of water from the stream. The volume of titrant required to change the color of the sample reflects the concentration of dissolved oxygen in the sample. Different test kits vary in the apparatus used to dispense the titrant. Options include:

• **Drop Bottle:** The titrant comes in a bottle with a dropper attached to the cap. The dropper is designed so that each drop dispenses the same, known volume of titrant.

• **Microburet:** The titrant is dispensed with a syringe-like apparatus that has a dropper opening rather than a needle. This method can be more accurate than the drop bottle. The microburet is marked with small volume increments, and the volume dispensed is measured by how far the syringe plunger is pushed along those marks.

• **Digital titrator:** This apparatus dispenses very small volumes of titrant, measured with a digital dial. This is the most accurate of the field test methods. It is also the most sensitive, measuring dissolved oxygen concentrations in increments of 0.1 mg/l.

Electronic Meters

Electronic meters can provide more precise readings of the dissolved oxygen concentration than field test kits. Their drawback is that they are much more expensive. The methods in this section are written for field test kits because we assume most Streamkeeper groups are not able to afford electronic meters.

If you conduct quality control checks on your field test kits as described in the procedure, your data can be reliable and useful. If you do use a meter, make sure you **calibrate** it frequently. You should receive instructions for calibrating your meter when you purchase it.

WHERE?

For baseline studies, select a site with a water sample typical of your overall stream reach and not representing a localized condition. For example, do not sample next to a discharge or drainage pipe. Try to choose a site free of suspended sediments, and deep enough so that you can collect your sample from beneath the surface of the water. Do not try to collect your sample in bubbly white water. The site you select for dissolved oxygen should be the same site you sample for other water quality parameters so you can compare the results of all tests with one another.

If you are trying to assess the impact of a particular discharge or other activity, you will need to consider localized conditions. Select a site just upstream from the suspected source of impact, a second site just downstream, and additional sites further downstream from the suspected source. Space out these additional downstream sites so you can determine the distance it takes your stream to recover from a suspected pollution problem.

WHEN?

For baseline information, we recommend that you measure dissolved oxygen at least monthly. For consistency, conduct your tests at the same time of the day each month.

For impact assessments, you should monitor the dissolved oxygen more frequently so that you can detect subtle changes on a regular basis.

If aquatic plants and algae are abundant in your stream, there may be a significant difference in dissolved oxygen concentration between early morning and late afternoon, especially in the summer. Conduct comparison tests to see if this is the case for your stream site. If so, make sure you take readings in the morning to pick up the daily minimum dissolved oxygen value.

HOW?

Before you head out to the field:

1. Read the instructions that accompany your dissolved oxygen test kit, including safety and material handling considerations.

2. For a quality control check, try out your test kit and make sure that all Streamkeepers are consistent in their methods.

3. Clean all containers/tubes thoroughly and rinse with distilled water.

In the field:

1. Follow the directions that accompany your dissolved oxygen test kit.

2. Use gloves while collecting the sample and conducting the test to avoid contamination of your water sample and to protect your hands. Wear glasses or goggles to protect your eyes.

3. Rinse testing containers/tubes a few times with the water you will sample.

4. Collect samples in the main current of the stream, away from the banks, and under the surface of the water, where homogeneous mixing of the water occurs.

5. Submerge the sample bottle into the water, allowing the water to fill it slowly. Be careful not to aerate the water sample or create turbulence during collection. Keeping the bottle submerged, slowly turn it upright to let it fill completely. Keep it submerged for 2-3 minutes. Cap the bottle while submerged and upright, and then remove from the stream. If any bubbles are present, try again. Bubbles signify that you have captured oxygen from the air in your sample, and this could result in a false, high reading.

6. Add the first two reagents to your sample immediately. This will "fix" the oxygen in your

sample so temperature changes and contact with the air will not affect the dissolved oxygen value. "Fixed" samples can be analyzed on site or stored for up to 8 hours if refrigerated and kept in the dark.

7. During the test, you will have to remove the stopper to add reagents and then replace it to mix. Take care not to add a bubble of air to your sample when you replace the stopper. If air bubbles appear in your sample at any time throughout the test, throw out your sample and start again.

8. We recommend you conduct three replicate tests at each sampling site. Record all values on Data Sheet 7, "Water Quality."

9. Be certain to carry out all waste products when you leave your stream site. The test products contain chemicals that could pollute your stream if left behind or poured back into the water. Refer to the material handling information provided with your test kits to find out about proper disposal methods.

NOW WHAT?

To report one result, average your values, discarding any **outlyers** (values that are way out of range and probably due to procedural error).

If you haven't done so already, find out what your state's dissolved oxygen standard is for your stream. Your stream's standard will depend on its classification, which is based on its "beneficial uses."

In general, dissolved oxygen levels less than 3 mg/l are stressful to most aquatic organisms. Cold water organisms such as salmon, pike, and trout usually require levels above 6 mg/l for growth and activity, while bass, catfish and carp prefer a little lower dissolved oxygen level (5 mg/l). Most fish will die at 1 - 2 mg/l. However, fish can move away from low dissolved oxygen areas. Water with low dissolved oxygen, from 2 - 0.5 mg/l are considered **hypoxic**; waters with less than 0.5 mg/l are **anoxic**.

For a quality control check, **calibrate** your test kit with a dissolved oxygen meter. Calibra-

tion refers to checking your equipment against a known outcome to determine the accuracy of the results. Plan to meet an agency scientist or water quality technician in the field so you can test samples of water taken from the exact same locations where meter readings are taken. See page 192 in Chapter 8, "Credible Data," for more information on accuracy and calibration procedures.

Because the temperature of a stream can vary daily, and even hourly, it is important to factor out the effect of temperature when analyzing the dissolved oxygen levels in a sample of water. This is achieved by considering the **saturation** value. Saturation is the maximum level of dissolved oxygen that would be present in the water at a specific temperature, in the absence of other influences. Table 10 on the following page lists the saturation values for dissolved oxygen in water ranging from 0 - 45 °C.

Once you know the temperature of the water in your stream (see "Field Procedure: Temperature" on page 178), you can use the oxygen saturation table to determine the maximum dissolved oxygen concentration. To compare this value with your actual measured dissolved oxygen result, you can calculate the **percent saturation**. Simply divide your measured result by the maximum value.

For example, if your stream temperature is 8 ° C, your maximum saturation value would be 11.83 mg/l. If your dissolved oxygen reading was 8.5 mg/l, your percent saturation would be 8.50/11.83 = 71.9 percent. Since a healthy stream is considered to be 90 - 100 percent saturated, your sample indicates that something else besides temperature is affecting oxygen levels adversely.

A stream as cold as 8° C should have more than 8.5 mg/l of dissolved oxygen. Suspended or dissolved solids or bacteria decomposition could be limiting this stream's ability to hold oxygen.

For a less precise but quick and easy method for obtaining a percent saturation value, use figure 7.2 on page 176. Simply use a straight edge to line up your measured values for temperature (top horizontal line) and dissolved oxygen (bottom horizontal line). The point where the straight edge intersects the diagonal line corresponds to the percent saturation value.

TABLE 10 CALCULATING PERCENT SATURATION

Some water quality standards are expressed in terms of percent saturation. To calculate % saturation of the sample:

1) Find the temperature of your water sample as measured in the field.

2) Find the maximum concentration of your sample at that temperature as given in the table below

3) To calculate the percent saturation: divide your actual dissolved oxygen measurement by the maximum concentration at the temperature of your sample.

Maximum Dissolved Oxygen Concentration

Temperature °C	Dissolved oxygen mg/l	Temperature °C	Dissolved oxygen mg/l
0	14,60	23	8.56
1	14.19	24	8.40
2	13.81	25	8.24
3	13.44	26	8.09
4	13.09	27	7.95
5	12.75	28	7.81
6	12.43	29	7.67
7	12.12	30	7.54
8	11.83	31	7.41
9	11.55	32	7.28
10	11.27	33	7.16
11	11.01	34	7.05
12	10.76	35	6.93
13	10.52	36	6.82
14	10.29	37	6.71
15	10.07	38	6.61
16	9.85	39	6.51
17	9.65	40	6.41
18	9.45	41	6.31
19	9.26	42	6.22
20	9.07	43	6.13
21	8.90	44	6.04
22	8.72	45	5.95

TABLE 12 DETERMINING PERCENT SATURATION
THE "QUICK AND EASY" METHOD

For a quick and easy determination of the percent saturation value for dissolved oxygen at a given temperature, use the saturation chart below. Pair up the mg/l of dissolved oxygen you measured and the temperature of the water in degrees C. Draw a straight line between the water temperature and the mg/l of dissolved oxygen. The percent saturation is the value where the line intercepts the saturation scale. Streams with a saturation value of 90% or above are considered healthy.

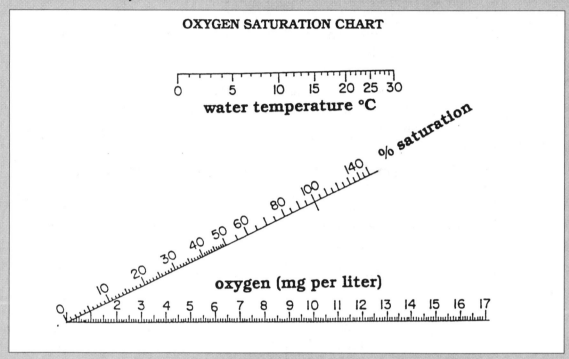

OXYGEN SATURATION CHART

TABLE 13 BIOLOGIC EFFECTS OF DECREASING DISSOLVED OXYGEN LEVELS ON SALMONIDS, OTHER FISH AND AQUATIC INVERTEBRATES

	Instream Dissolved Oxygen		Instream Dissolved Oxygen
I Salmonid Waters			
A. Embryo and larval stages			
No production impairment	11	Slight production impairment	5.5
Slight production impairment	9	Moderate production impairment	5
Moderate production impairment	8	Severe production impairment	4.5
Severe Production impairment	7	Limit to avoid acute mortality	4
Limit to avoid acute mortality	6		
		B. Other life stages	
B. Other life stages		No production impairment	6
No production impairment	8	Slight production impairment	5
Slight production impairment	6	Moderate production impairment	4
Moderate production impairment	5	Severe production impairment	3.5
Severe production impairment	4	Limit to avoid acute mortality	3
Limit to avoid acute mortality	3		
		III Invertebrates	
II Non-salmonid waters		No production impairment	8
A. Early life stages		Moderate production impairment	5
No production impairment	6.5	Limit to avoid acute mortality	4

From *Monitoring Guidelines to Evaluate Effects of Forestry Activity on Streams in the Pacific Northwest and Alaska.* U.S. Environmental Protection Agency, Region 10, Seattle, WA.

The Streamkeeper's Field Guide

TEMPERATURE

Water temperature is a controlling factor for aquatic life: it controls the rate of metabolic activities, reproductive activities and therefore, life cycles. Most aquatic organisms are **cold blooded,** which means they cannot regulate their own body temperatures. They assume a temperature similar to the surrounding water and carry out their metabolic activities within this range.

Cold blooded organisms are finely adapted to a specific temperature regime and deviations out of the usual range may cause problems. If stream temperatures increase, decrease, or fluctuate <u>too</u> widely, metabolic activities may speed up, slow down, malfunction, or stop altogether.

As discussed earlier in this chapter, temperature affects the concentration of dissolved oxygen in a water body. Some organisms, like salmonids, require high oxygen and thus can only live in environments that have cool temperatures and high oxygen concentrations. Temperature also influences the activity of toxic chemicals, par-asites, diseases, and the sensitivity of aquatic organisms to the chemicals and diseases.

There are many factors which influence the temperature of stream water. As you might expect, temperature fluctuates seasonally, being cooler in winter and warmer in summer. Temperature may also fluctuate daily and hourly, especially in small streams in the summer, being cooler in the morning and warmer by the end of the day.

However, there are factors that can buffer the effect of summer heat and keep stream temperatures cool. For example, springs discharging cool groundwater into streams keep water temperatures down. The overhanging canopy of streamside vegetation which provides shade and thus keeps stream water cool.

Water temperature is also influenced by the quantity and velocity of stream flow. The sun has much less effect in warming the waters of streams with greater and swifter flows than of streams with smaller, slower flows.

In addition, human activities can alter the temperature regime of a stream. Water released into streams from power plants and industrial facilities, and return flows from irrigation systems can increase the temperature of a stream significantly.

Water releases from dams can also alter temperature regimes. Blocking the flow of water in a stream with a dam creates ponds or lakes that often have warm surface waters and cold bottom waters. Depending on what layer is released from the dam, temperatures downstream may increase or decrease significantly downstream. Bottom water releases also pose an oxygen supply problem for aquatic organisms, as explained in the earlier section on dissolved oxygen.

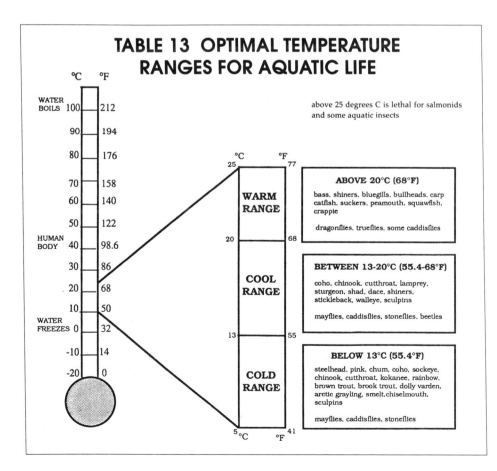

TABLE 13 OPTIMAL TEMPERATURE RANGES FOR AQUATIC LIFE

above 25 degrees C is lethal for salmonids and some aquatic insects

WARM RANGE

ABOVE 20°C (68°F)

bass, shiners, bluegills, bullheads, carp catfish, suckers, peamouth, squawfish, crappie

dragonflies, trueflies, some caddisflies

COOL RANGE

BETWEEN 13-20°C (55.4-68°F)

coho, chinook, cutthroat, lamprey, sturgeon, shad, dace, shiners, stickleback, walleye, sculpins

mayflies, caddisflies, stoneflies, beetles

COLD RANGE

BELOW 13°C (55.4°F)

steelhead, pink, chum, coho, sockeye, chinook, cutthroat, kokanee, rainbow, brown trout, brook trout, dolly varden, arctic grayling, smelt, chiselmouth, sculpins

mayflies, caddisflies, stoneflies

FIELD PROCEDURE: TEMPERATURE

EQUIPMENT NEEDED

calibrated Celsius thermometer

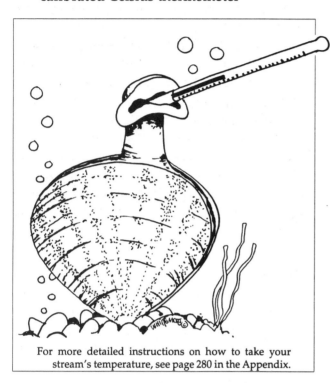

For more detailed instructions on how to take your stream's temperature, see page 280 in the Appendix.

WHAT?

This procedure provides instructions on how to measure the temperature of your stream.

WHERE?

Measure temperature at the same locations where you measure pH and dissolved oxygen, close to where you collect macroinvertebrates.

WHEN?

We recommend that you monitor temperature on a monthly basis, at the same time of the day each month. During the summer, you may also want to monitor temperature fluctuations from morning to evening.

HOW?

1. Tie a string to the top end of your thermometer and lower it to about four inches below the surface. Do not hold the thermometer itself while taking your reading.

2. Keep the thermometer in the water for approximately two minutes, or until a constant reading is obtained.

3. Record the value on Data Sheet 7, "Water Quality."

NOW WHAT?

Compare your results to state and federal standards for various water uses.

Compare your temperature reading with your dissolved oxygen test results to find the percent saturation value. (See "Field Procedure: Dissolved Oxygen").

Look at the information on the temperature ranges for aquatic life provided in figure 7.3 to determine what types of life your stream can support. Are your results consistent with your watershed inventory information and stream survey data? Is the temperature level preventing any populations from thriving in your stream?

THE TEMPERTURE TEST

TOO COLD.

A LITTLE COLD.

EARLY FUZZY LOGIC

TESTING FOR OTHER PARAMETERS

As you get to know more about your stream's water quality from testing pH, dissolved oxygen, and temperature, you may want to look at other water quality parameters. A few of the commonly monitored parameters are discussed below. For some of these parameters, testing involves use of expensive meters or more complicated procedures, such as incubation, heating, or toxic waste products that require special disposal. For more information, consult the test kit suppliers or the water quality references listed in the reference section in the back of this book.

Alkalinity

The **alkalinity**, or the **buffering capacity** of a stream refers to how well it can neutralize acidic pollution and resist changes in pH. Alkalinity measures the amount of alkaline compounds in the water, such as carbonates, bicarbonates, and hydroxides. These compounds are natural buffers that can remove excess hydrogen (H+) ions.

As increasing amounts of acid are added to a body of water, its buffering capacity is consumed. If surrounding soils and rocks are alkaline, they may eventually restore the buffering capacity, but a temporary decrease in alkalinity can allow the pH to drop to harmful levels.

Measuring alkalinity involves a **titration**. An acid of known strength (the titrant) is added to a specific volume of a treated sample of water from the stream. The volume of acid required to bring the sample to a specific pH level reflects the alkalinity of the sample. The pH end point is indicated by a color change.

A total alkalinity of 100 - 200 mg/l will stabilize the pH level in a stream. Levels between 20 and 200 mg/l are typically found in fresh water.

Biochemical Oxygen Demand (BOD)

The **Biochemical Oxygen Demand (BOD)** is the amount of oxygen consumed by bacteria in the decomposition of organic material. It also includes the oxygen required for the oxidation of various chemicals in the water, such as sulfides, ferrous iron, and ammonia.

While a dissolved oxygen test tells you how much oxygen is available, a BOD test tells you how much oxygen is being consumed. Thus if you find a dissolved oxygen problem, a BOD test will help you determine potential problem sources.

BOD is determined by measuring the dissolved oxygen level in a freshly collected sample and comparing it to the dissolved oxygen level in a sample that was collected at the same time but incubated under specific conditions for a certain number of days. The difference in the oxygen readings between the two samples is the BOD, recorded in units of mg/l.

Unpolluted, natural waters should have a BOD of 5 mg/l or less. Raw sewage may have BOD levels ranging from 150 - 300 mg/l. Levels required in the discharge permits for wastewater treatment plants usually range from 8 to 150 mg/l.

Nutrients

Nutrients such as nitrogen and phosphorus are required by all organisms for basic processes of life, growth and reproduction. Various forms of nitrogen and phosphorus naturally occur in stream water.

In excessive amounts, nutrients overstimulate the growth of aquatic plants and algae. The vegetation can clog waterways, and when it dies and is decomposed by bacteria, dissolved oxygen levels can drop dramatically. This lack of oxygen can adversely affect fish and aquatic invertebrates, leading to decreased community diversity.

Sources of excess nutrients include: fertilizers applied to agricultural fields, timbered areas, gardens, and lawns; poorly maintained septic systems, sewage treatment plants; industrial effluent; pet, livestock, and other animal wastes; and detergents.

Nitrogen

Nitrogen occurs in natural waters in various forms, including nitrate (NO_3), nitrite (NO_2), and ammonia (NH_3). Nitrate is the most common form tested. Test results are usually expressed as **nitrate-nitrogen** (NO_3-N), which simply means nitrogen in the form of nitrate. Ammonia is the least stable form of nitrogen and thus difficult to measure accurately. Nitrite is less stable and usually present in much lower amounts than nitrate.

Measuring nitrate requires a chemical reaction that yields cadmium, a toxic metal that requires special disposal. The treated sample can then be analyzed **colorimetrically**, which means the color of the sample reflects the concentration of the parameter being measured. The darker the color, the greater the concentration of the parameter.

Analyzing color can be done visually, with the human eye, using a set of reference colors for comparison. However, this is a fairly crude method that cannot detect small amounts of nitrate or small incremental changes. An electronic **colorimeter** can analyze color more accurately, precisely, and with greater sensitivity. A colorimeter is a device with a light source, a photodetector, and an analog or digital readout. It tells you the concentration of your parameter based on how much light is absorbed by your sample. (Darker color absorbs more light).

The national drinking water standard for nitrate-nitrogen in the United States is 10 mg/l. However, polluted waters generally have a nitrate-nitrogen level below 1 mg/l.

Phosphorus

Phophorous usually occurs in nature as **phosphate**, which is a phosphorous atom combined with four oxygen atoms (PO_4^{-3}). Phosphate that is bound to plant or animal tissue is known as **organic phosphate**. Phosphate that is not associated with organic material is known as **inorganic phosphate**. Both forms are present in aquatic systems and may be either dissolved in the water or suspended (attached to particles in the water column).

Inorganic phosphate is often referred to as **orthophosphate** or **reactive phosphorous**. It is the form most readily available to plants, and thus may be the most useful indicator of immediate potential problems with excessive plant and algal growth. It is also the easiest to measure.

Testing for **total phosphorous** (both inorganic and organic phosphate) provides you with a more complete measure of all the phosphorous that is actually in the water. However, it requires "digestion," or heating the sample to break down organic phosphate so it can be measured. This can be difficult to conduct safely and accurately in the field.

You can also measure **dissolved phosphorous**, to determine how much of the phosphorous in your stream's water is dissolved versus **insoluble** (suspended in the water column). This is helpful in determining the source of phosphorous. Insoluble phosphorous comes from erosion, manure, or sewage, while dissolved phosphorous comes from chemical fertilizer or septic system leachate.

Like the total phosphorous test, the dissolved phosphorous test is also difficult to conduct accurately in the field. Measuring dissolved phosphorous requires filtration. Insoluble phosphorous is filtered from the sample so that dissolved phosphorous can be measured.

Once the various phosphorous samples are treated, they are analyzed colorimetrically, like nitrate. Phosphorous is often the limiting nutrient for plant growth, meaning it is in short supply relative to nitrogen. Very small amounts of phosphorous - even as low as 0.01 mg/l - can have a significant impact on the plant growth in a stream, especially slower moving areas. If you wish to detect such small amounts of phosphorous, you need to use a colorimeter to achieve that level of sensitivity.

Bacteria

Human and animal wastes carried to stream systems are sources of pathogenic, or disease-causing, bacteria and viruses. The disease causing organisms are accompanied by other common types of non-pathogenic bacteria found in animal intestines, such as fecal coliform bacteria, enterococci bacteria, and escherichia coli (E. coli) bacteria.

Fecal coliform, enterococci, and E. coli bacteria are not usually disease-causing agents themselves. However, high concentrations suggest the presence of disease-causing organisms. Fecal coliform, enterococci, and E. coli bacteria are used as indicator organisms; they indicate the probability of finding pathogenic organisms in a stream.

Concentrations of pathogenic bacteria have been correlated with increased concentrations of sediment. The organisms attach to sediment particles and hitch a free ride, escaping their invertebrate predators, absorbing nutrients from the sediment and extending their brief lifetimes.

Other sources contributing to the occurrence of pathogenic bacteria in our streams are septic tank failure, poor pasture management and animal keeping practices, pet waste, urban runoff, and sewage from storm water overflows.

To measure indicator bacteria, water samples must be collected in sterilized containers. The samples are forced through a filter and incubated at a specific temperature for a certain amount of time. The resulting colonies that form during incubation are counted and recorded as the number of **colony producing units** per 100 ml of water (CPU/100 ml).

Safe drinking water should have fecal coliform counts of 0 CPU/100 ml. For safe swimming water, less than 200 CPU/100 ml is usually desirable, and for boating, less than 1000 CPU is desirable. However, local, state, and federal standards may allow higher CPU levels.

Conductivity

Conductivity is a measure of how well water can pass an electrical current. It is an indirect measure of the presence of inorganic **dissolved solids** such as chloride, nitrate, sulfate, phosphate, sodium, magnesium, calcium, iron, and aluminum. The presence of these substances increases the conductivity of a body of water. They help to conduct electricity because they are negatively or positively charged ions when dissolved in water. Organic substances like oil, alcohol, and sugar do not conduct electricity very well, and thus have a low conductivity in water.

Dissolved solids are essential ingredients for aquatic life. They regulate the flow of water in and out of organisms' cells, and are building blocks of the molecules necessary for life. A high concentration of dissolved solids, however, can cause water balance problems for aquatic organisms and decreased dissolved oxygen levels, as discussed in the section on dissolved oxygen in this chapter. Too many dissolved solids can also lead to poor tasting drinking water with laxative effects. **UH OH!**

The concentration of dissolved solids, or the conductivity of a stream, is affected by the bedrock and soil make up of the watershed. Conductivity tends to be low in areas where soil and bedrock is composed mainly of **inert** materials (materials which do not dissolve into ions when mixed with water). Conductivity tends to be higher in areas with clay soils, because clay tends to contain substances that do dissolve into ions when washed into a stream.

Conductivity is also affected by human influences. Failing sewage systems and agricultural runoff tend to raise conductivity due to the presence of phosphate and nitrate. Oil spills tend to lower conductivity. Industrial discharges may raise or lower the conductivity, depending on their chemical makeup.

Conductivity is measured in **micromhos** per centimeter. Abbreviated μmho, a μmho is a unit of current, or the flow of electricity. Distilled water has a conductivity ranging from 0.5 - 3 μmhos/

cm. Generally, streams supporting healthy fish populations range between 150 - 500 µmhos/cm. Some industrial waters can range as high as 10,000 µmhos/cm.

To measure conductivity, an electronic meter is required. No reagents are needed. It can be measured easily either in the field or in a lab. If you do not have a conductivity meter, you can measure the concentration of **dissolved solids** directly using the method described under "Total Solids," below.

Turbidity

Turbidity is a measure of the cloudiness of water. Cloudiness is caused by **suspended solids** (mainly soil particles) and **plankton** (microscopic plants and animals) that are suspended in the water column. Moderately low levels of turbidity may indicate a healthy, well-functioning ecosystem, with moderate amounts of plankton present to fuel the food chain.

However, higher levels of turbidity pose several problems for stream systems. Turbidity blocks out the light needed by submerged aquatic vegetation. It also can raise surface water temperatures above normal because suspended particles near the surface facilitate the absorption of heat from sunlight.

Suspended soil particles may carry nutrients, pesticides, and other pollutants throughout a stream system, and they can bury eggs and benthic critters when they settle. Turbid waters may also be low in dissolved oxygen, as discussed in the dissolved oxygen section in this chapter. High turbidity may result from sediment bearing runoff, or nutrient inputs that cause plankton blooms.

Turbidity in slow moving, deep waters is easily measured with an inexpensive, often homemade device known as a **Secchi disk**. A Secchi disk is a 20 cm disk with black and white quadrants attached to a line. The disk is lowered into the water until it just disappears from sight. The depth at which the disk disappears, or the **Secchi depth**, is recorded in meters.

A Secchi disk does not work in shallow, fast moving streams. The best instrument for measuring stream turbidity is a **nephelometer**. A nephelometer is an electronic meter that measures the amount of light scattered by a sample of water. The greater the amount of suspended particles, the greater the intensity of scattered light. Turbidity measured using a nephelometer is recorded in **Nephelometric Turbidity Units (NTU's)**.

A lower cost method for measuring turbidity in streams is available, but it can only be used for measuring relatively high levels and large changes in turbidity. The method involves evaluating the fuzziness of a mark at the bottom of a clear tube when a sample of water is poured into the tube. The result is recorded in **Jackson Turbidity Units (JTU's)**, named after the inventor of the prototype of this method.

If you do not have a nephelometer and a Jackson tube is not sensitive enough for your purposes, you can measure the concentration of **suspended solids** directly using the method described in the following section entitled "Total Solids." Be aware that suspended solids is only one factor contributing to overall turbidity; measuring suspended solids does not measure plankton.

Total Solids

Total solids is a measure of the suspended and dissolved solids in a body of water. Thus, it is related to both conductivity and turbidity. If you do not have a nephelometer or conductivity meter, but you have access to a lab with a sensitive scale and a drying oven, this method may be a viable alternative for you.

To measure total suspended and dissolved solids, a sample of water is placed in a drying oven to evaporate the water, leaving the solids. The dry sample bottle is then weighed with a sensitive scale. To measure dissolved solids, the sample is filtered (to remove suspended solids) before it is dried and weighed. To calculate the suspended solids, the weight of the dissolved solids is subtracted from the total solids.

DATA SHEET 7 WATER QUALITY

Name_____ Group_____ Date_____ Time_____

Stream Name_____ Reach Name/#_____

Site Description: (site name, landmarks, etc.)_____

Site Location: (lat & long, transect #, and/or rivermile) _____

Driving and/or Hiking Directions_____

Weather Conditions: ❏ Clear ❏ Cloudy ❏ Rain ❏ Other _____

Air temperature: _____° ___ Recent weather trends:_____

Quantitative Water Quality

pH	Dissolved O_2	Temperature
1_____	1_____	_____° ___
2_____	2_____	
3_____	3_____	_____% Dissolved Oxygen Saturation
mean_____	mean _____	

Qualitative Water Quality

Water Appearance	Stream Bed Coating	Odor
❏ Scum	❏ Orange to red	❏ Rotten egg
❏ Foam	❏ Yellowish	❏ Musky
❏ Muddy	❏ Black	❏ Acrid
❏ Milky	❏ Brown	❏ Chlorine
❏ Clear	❏ None	❏ None
❏ Oily sheen		
❏ Brownish		
Other_____	Other_____	Other_____

Other Tests:

Comments:

PART THREE

Effective Information

## CHAPTER EIGHT

Credible Data

Now you know what a watershed is. You know how to investigate all of the available information on your watershed. You have adopted a stream. You know how to conduct a watershed inventory, and how to map a stream reach. You can monitor the physical, biological, and chemical characteristics of that stream. And you are getting geared up to do just that.

Undoubtedly you will want to use the information you collect to teach others about the ecology of your stream and encourage them to join you in becoming a steward of your watershed. You may want to use your data to document water resource problems and influence local officials making land use decisions. Your data could become the information needed to design a fish ladder or create wildlife habitat. It may become part of a database used by local, state, and federal government agencies. Or, possibly, you might want to use your data in court.

"Data" Defined

Data is "facts or information to be used as a basis for discussing or deciding something."

Oxford American Dictionary

In order to use your data, however, you should have a high degree of comfort that the information you collect is considered credible. If your data is not credible, you may not be able to use it for the purposes you intended. The best way to ensure credible data is to develop a thorough **Quality Assurance Plan**. A Quality Assurance (QA) Plan provides a blueprint for designing and evaluating your watershed inventory and stream monitoring program to ensure

that your data is of the quality needed to meet your goals. A QA Plan also spells out specific **Quality Control** (QC) steps that you take while collecting and analyzing information to ensure that your data is credible.

A Quality Assurance Plan should answer the following questions:

- Who are you and what is your expertise?
- Why are you monitoring?
- What training do you have?
- What monitoring protocols did you use?
- What equipment did you use?
- Was your equipment calibrated?
- Who did the actual tests, and what kind of training did they have?
- Were your testing materials within expiration parameters?
- When and where did you collect your samples?
- How did you record the results of your tests?
- How did you analyze your sample results?
- How did you evaluate the accuracy of your results?
- How accurate were your visual observations?

Quality Assurance and Quality Control can occur at different levels of detail. Euhania Hairston, Principal of Seattle's Martin Luther King Elementary School, says:

"Our first graders can take the temperature of a stream as well as a Ph.D., but we have to help them in averaging the readings of several locations. . . the third graders are able to grasp the need to check their math to make sure that their flow readings are accurate...the fifth graders are beginning to understand the need to repeat chem-

BUREAUCRAT LURE

check the student's accuracy and validate their findings."

By describing standardized information (data) gathering procedures in writing, the Martin Luther King teachers have in effect created a lesson plan that they can use into the future. That lesson plan is also an elementary Quality Assurance Plan. By observing how the students are collecting data, and checking the results, the teachers are providing Quality Control.

As your monitoring skills are honed, your procedures and methods may become more complex, and your Quality Assurance Plan will become more sophisticated. We recommend you get a three ring binder notebook for your QA Plan and continue to develop it as your inventory and monitoring program continues to grow and expand.

Editor's Note: When you are tackling a tough problem involving diverse interest groups, make sure your Quality Assurance plan will stand up to outside scrutiny.

DESIGNING A QUALITY ASSURANCE PLAN

So, what are the required elements of a thorough Quality Assurance plan? The U.S. Environmental Protection Agency (EPA) recently developed 24 elements required of all QA Plans for state monitoring programs receiving EPA monitoring funds. Most successful volunteer monitoring programs across the U.S. use the same process.

The following is our **translation** of the EPA's Quality Assurance Plan elements. Assisting in the translation was Sam Stribling, from Tetra Tech, Inc. an environmental consulting firm in Owings Mills, Maryland. Sam has had the dubious honor of working on these QA procedures during the last four years for the EPA.

Note that there is nothing sacred about the format presented here. As you read through each element, you may detect a certain amount of overlap and redundancy, then decide to merge some of those QA elements for your monitoring plan. That's okay. **Just make sure that the Quality Assurance Plan you develop answers all of the questions listed on page 186.** Your QA Plan should include all of the information needed to make anyone who uses your data comfortable that it is reliable information.

QA May Not Be Fun, But It's Important

We recognize that the paperwork part of being a Streamkeeper is not the most entertaining Streamkeeper activity, but it can be the most important. As Massachusetts Resources Research Center staff scientist Mark Mattson puts it:

"Let's face it .... quality assurance is probably the least glamorous aspect of any environmental monitoring program. But the scientist, public officials, and members of the public who may use your data have a right to know how accurate and reproducible those data are, and a strong quality assurance program is the only way to convince these people that your data are worth looking at."

---

1. Title Page

Give your project a title, e.g., "The Rocky Creek Watershed Monitoring Program." List the name of the organization in charge of the project, and the name, address, and telephone number of the person considered the manager of the project. You should also list any funding organizations that may be supporting the project. Make sure that you keep addresses and telephone numbers current.

If you are collecting information subject to review and approval by a scientific or governmental body or other entity, make sure they review and approve your plan and sign and date a signature block at the bottom of the title page.

On the "who's in charge" suggestion above, there is no doubt that an individual Streamkeeper can be effective. Our experience leads us to conclude, however, that a well organized group of Streamkeepers will be able to accomplish more than any single individual. And, in a public forum, representatives of organizations with large numbers usually have more clout than individuals. Additionally, you will likely have more credibility if you are part of a group with an official sounding name than you will if you are working alone.

Picture yourself as part of a decision making body receiving testimony on a water resource issue. Think about what sounds more credible: "I'm Steve Jones, and I have data that indicates that there is a major water quality problem in Crystal Creek" or "I'm Steve Jones, Project Manager for the Crystal Creek Protection Association, an organization of 30 trained volunteers which has been monitoring Crystal Creek in coordination with the State Department of Ecology during the last five years. We have noted significant deterioration in the water quality of Crystal Creek." We assure you that the second Mr. Jones will have greater credibility right off the bat.

With this in mind, we recommend that you include (perhaps on the flip side of your title page) a brief statement providing background on the organization in charge and the qualifications of the Project Manager.

JULIET QUESTIONS THE TAXANOMIC SYSTEM

WHAT'S IN A NAME? WHAT'S IN A NAME?!

2. Table of Contents

A table of contents is an obvious but often over-looked element of a well written plan that should enable you, your fellow Streamkeepers, and anyone else to pick out any element of your QA plan. Be sure to amend the table of contents as your plan grows.

3. Distribution List

Identify all parties with whom you intend to share your information. Develop a list of key contacts that includes addresses and telephone numbers. Be sure to send all parties any changes that you make in your QA plan.

Who should you include on your distribution list? Identify your target audience. Is it your watershed neighbors? Community groups? Elected officials? Agency scientists? Since each target audience will likely respond differently to your data, include a summary of the level of data that each audience is likely to understand and any concerns they may have. Identify the mode of communication that would best reach each audience, i.e., phone call, personal meeting, letter, public hearing, rubber hammer (just kid-

ding). Chapter 9, "Presenting Your Data," focuses on this topic in greater detail.

You can add credibility to your program by including members of your target audience in the development of your QA plan. If you can swing this, you may even be in a good position to turn those that you perceive as "the bad guys" into "good guys."

4. Project Organization

The next step is to decide what roles you and your fellow Streamkeepers will have in your monitoring program. The project organization section of your QA plan should list the major roles that your project requires, a brief description of the duties involved for each role, and the person chosen to be in charge of each task. While this all sounds very bureaucratic, it will make your job easier in the long run. Including such "job descriptions" in your QA plan will help you identify Streamkeepers best suited for each role as well as any training that may be needed.

TARGET AUDIENCE

DATA

Some key roles that your Streamkeeper monitoring project may include are listed below:

- **Project Manager**: responsible for collecting quarterly and annual reports from monitoring coordinators, financial and administrative tasks, and other "fun" stuff.

- **Monitoring Coordinator**: responsible for ensuring all monitoring teams collect data on schedule, quality control check on data provided by monitoring teams.

- **Site Data Collectors**: responsible for collecting all data from specific sites according to procedures outlined in the QA plan.

- **Site Data Reporters**: responsible for summarizing all physical, biological and chemical monitoring information from specific sites.

- **Project Data Analyzers**: responsible for analyzing data gathered from all sites in the project and preparing reports on findings.

5. Problem Definition / Background

The problem definition is the specific problem that you'd like to see resolved, or the specific decision that needs to be made. In this section, include background information to provide a historical perspective of the situation. Include regulatory or alleged pollution or habitat degradation issues that have led you to identify the need for your monitoring project. You should also include references to previous studies that pertain to the problem.

For example, you may find that a new shopping center is being constructed in your stream's headwater wetlands. In this case, it would be helpful to provide information regarding the amount of impervious surface that will be placed on site, the volume of water that will be displaced by the planned wetland filling, and any available information on the function, value, and ecological health of the wetland and stream. In other words, include in your QA plan the information you gathered in your watershed inventory that is relevant to the problem at hand.

6. Project Description

The purpose of the project description is to define the specific objectives of your monitoring project and describe how the project will be designed to obtain the information needed to accomplish these objectives. A thorough project description includes a general overview, a description of how you've designed your study, and a timetable.

In the general overview, describe the area that you will be evaluating and the overall approach you intend to use to tackle the problem identified in section I-5. Recall from Chapter 3 that there are different approaches to monitoring that are used to address different types of problems.

For example, you may decide to conduct a **baseline study** if no watershed inventory and stream monitoring data exists in your watershed. Or, as in the example described in I-5, you may decide to conduct an **impact assessment** of the new shopping center slated for your stream's headwater wetlands. Or you may decide to conduct **compliance monitoring** to determine if the runoff or effluent flowing into your stream from some land use or industrial discharge meets your state or provincial water quality standards.

No matter what approach you choose, be sure to describe how you anticipate the data will be used to answer questions and make decisions.

In the study design description, list all planned data collection parameters, the number and types of samples you plan to collect, and the equipment you will use. Also describe how you will evaluate the data. Remember that some "parameters" you may want to include with your monitoring are not necessarily physical, chemical, or biological. You may decide to monitor new development activities, or even public perception of the health of the watershed.

Include a project timetable that lists the beginning and ending dates for the overall project and specific activities within the project. List potential constraints that may require flexibility, such as seasonal variations, sampling logistics, or site access issues. See Chapter 3 for a discussion of when and how long to monitor. Also, include clear directions to your monitoring locations, as described in "Documenting Locations," page 74.

7. Data Quality Objectives

As a Streamkeeper, you want your data to be viewed as credible by all audiences. In order to achieve this goal, you'll need to make sure the data you collect on each parameter you measure meets certain **data quality objectives** ("DQO's" in EPA lingo).

In this section of your QA plan, outline your DQO's by stating clearly how **accurate**, **precise**, **representative**, **comparable**, and **complete** you expect your data to be. For each of these qualities (which are defined in the following sections), you will have separate DQO's. In some cases, your DQO's will be expressed as numerical values; in other cases, your DQO's will be expressed in narrative form.

A. Accuracy and Precision

Sergio Huerta, M.D., of the State of Delaware Department of Natural Resources and Environmental Control suggests that to understand accuracy and precision you should, "envision yourself at a target range. The target is in front of you. You are given fifteen shots and are asked to shoot at the target in three rounds with each round having five shots."

"In the first round your (five) shots hit all over the target. These shots are **neither accurate nor precise**."

Neither Precise nor Accurate

"In the second round, the five shots are closely grouped, but they miss the bulls eye. These are **not accurate but they are precise**."

Precise, but _not_ Accurate

"In the third round, the shots are closely grouped in the bulls eye; these are **accurate and precise**."

Precise _and_ Accurate

Finally, if you repeat this third round performance many times over, the shots are considered to be **accurate, precise, and reliable**."

Precise, Accurate, and _Reliable_

Sergio goes on to say, "in environmental monitoring, the data collected (e.g., dissolved oxygen, pH, physical and biological observations, etc.) must be determined to be accurate, precise, and reliable before it can be said that it realistically describes conditions or problems in nature."

Human error can have a significant effect on how precise your readings are, while the quality of your measuring and data collection tools will determine your accuracy. No matter how good the quality of the equipment, if Streamkeepers are not using careful, consistent methods, precision will suffer.

On the other hand, extremely precise methods will not yield accurate results if the equipment is faulty or not calibrated properly. For example, temperature measurements will be precise if they are taken the same way by all Streamkeepers, and accurate if the thermometers used are calibrated with a certified thermometer.

Data Quality Objectives Defined

• **Accuracy**: How close your results are to the true value. This term generally refers to the equipment or procedures used gathering data. For example, distance measurements will be more accurate using a tape measure than pacing.

• **Precision**: How well you can reproduce your data result using the same equipment and sample. This term generally reflects human error rather than equipment failure, in other words, "are you following your inventory and monitoring procedure directions to the letter?"

Ensuring Accuracy - Calibrating Your Equipment

To ensure accuracy, maintain your equipment properly and **calibrate** it often. Calibration refers to performing a test with an expected outcome, comparing your result with the expected outcome, and then adjusting your equipment or results appropriately.

For example, to calibrate a pH test kit, you could **test buffer solutions** of known pH values. Try this with three buffers, such as one at pH 4, one at pH 7, and one at pH 10. If your results are different from the expected pH value, record the difference between your results and the expected values. Your accuracy would be plus or minus the greatest difference that occurred between one of the expected values and your result. You can express this as "+/- N," where N equals the greatest difference.

You can also **calibrate your test kits to meters**. Assuming that the meter is reading the

"true" value (it should also be calibrated), you can compare your test kit result to the meter reading to see how accurate your method is.

To calibrate a test kit to a meter, we recommend you conduct four paired tests at your stream. Collect four different water samples, one for each paired test. For each paired test, make sure you use water from the same sampling container to do both the test kit analysis and the meter reading.

Calculating Accuracy

If your test kit results do not vary from the meter readings, your level of accuracy is 100%. If they do vary, you can determine your accuracy as follows:

1. Calculate the **mean** ($\overline{X}$) kit values ($\overline{X}_{kit}$) and the mean of the meter values ($\overline{X}_{meter}$). A mean is calculated by summing all results and dividing the sum by the total number of results.

2. Calculate the **standard deviation** (S) of the mean of the test kit values ($\overline{X}_{kit}$) and the standard deviation of the mean of the meter values ($\overline{X}_{meter}$) using the formula below. A standard deviation is simply a mathematical measure of how dispersed the values are that were used to calculate a mean.

Standard Deviation or S

$$ S = \sqrt{\sum_{i=1}^{n} \frac{(x_1 - \overline{X})^2}{n-1}} $$

where $\overline{X}$ = mean of the measured values for each test
X_i = measured value of each test
and   n = number of tests

The "Σ" symbol is simply a fancy way to say "take the sum of." The expression

$$ \sum_{i=1}^{n} $$

where n = the number of tests, refers to the sum over all the tests.

Thus if you did four tests with the test kit, you would calculate the standard deviation of X_{kit} as follows:

Standard deviation of $\overline{X}_{kit} = S_{kit} =$

$$\frac{(X_1 - \overline{X}_{kit})^2}{3} + \frac{(X_2 - \overline{X}_{kit})^2}{3} + \frac{(X_3 - \overline{X}_{kit})^2}{3} + \frac{(X_4 - \overline{X}_{kit})^2}{3}$$

where X_1, X_2, X_3, and X_4 are the four values you obtained with the test kit

You would calculate the standard deviation of $\overline{X}_{meter}$ (S_{meter}) the same way, substituting the four values you obtained with the meter as X_1, X_2, X_3, and X_4.

3. Calculate the **relative percent difference** (RPD) between S_{kit} and S_{meter}. The relative percent difference is a simple measure of the difference between two values. You can calculate it using the following formula:

Relative Percent Difference or RPD

$$RPD = \frac{(C_1 - C_2) \times 100}{(C_1 + C_2) / 2}$$

where C_1 = the larger of the two values
and C_2 = the smaller of the two values

Thus if S_{kit} was greater than S_{meter}, C_1 would be S_{kit} and C_2 would be S_{meter}.

You can report the RPD value as your level of accuracy. In general, a 90 - 95% RPD is a desirable accuracy DQO. If your level of accuracy is below 90%, you have to decide if you are satisfied with this as an accuracy DQO or if you want to try another test kit or measurement method.

Before you give up on a particular test kit or method, you may want to repeat this procedure one or more times to ensure that the low level of accuracy is not due to sampling errors that can be corrected with more careful field protocol.

Calibrating Other Field Methods

You can calibrate your flow measurement method to a flow meter and use step 3 above to calculate the level of accuracy. Instead of doing four paired tests, simply measure flow using your float method in the same spot as a meter is used. To determine your accuracy, calculate the RPD using the float method result and the meter result as C_1 and C_2.

You can calibrate your thermometers using a National Institute of Standards and Technology (NIST) thermometer. You may be able to borrow one of these high tech thermometers from your local health district or state environmental quality department.

Calibrating your bug method is a little different. It does not involve calculating an accuracy value or listing a numerical DQO. It involves issues like ensuring that your mesh size and net dimensions are appropriate. Your QA plan should state the mesh size of your net and any other standards you are using. If the mesh size of your net is unknown, staple a sample of the mesh in your QA plan.

Ensuring Precision - Replicate Sampling

To ensure precision, make sure your sampling procedures are consistent and repeatable. Thus all participating Streamkeepers should be trained in the same methods and use the same types of equipment. For example, all groups should sample the same size area of the stream

ANOTHER CASE OF REPLICATE SAMPLING

bottom when collecting bugs and use the same mesh size for their nets. Streamkeepers should be interpreting color codes for a pH test the same way and use the same types of reagents. All groups should use the same type of float for velocity measurements, if possible (make sure all your "rubber ducks" are purchased from the same manufacturer).

In order to determine your level of precision, you can collect and analyze a series of **replicate** samples. Replicate samples are two or more samples taken from the same place at the same time. They are expected to have the same, or very similar, outcomes. Thus the more similar your outcomes from different replicates are, the more precise your methods.

For example, your Streamkeeper group may use five different thermometers to take readings

at five different sites at the same time of day. You can check the precision of these five thermometers by putting them together in a bucket of water to see how variable their readings are.

Precision can be expressed as the range of values obtained from your measurements. For example, if you obtained temperature readings of 15, 16, 18, 14, and 15 °C from the five replicates, your precision would be 14 - 18 °C.

Calculating Precision

Precision can also be calculated quantitatively to be expressed as a single value. If you are only using two replicates, relative percent difference can be used to express precision. Calculate relative percent difference as described on page 193, using the replicate that yielded the larger result as C_1 and the other as C_2.

If you are using three or more replicates, you can express precision as the **relative standard deviation** (RSD). To calculate relative standard deviation, use the following formula:

Relative Standard Deviation or RSD

$$RSD = \frac{(S)}{\overline{X}} \times 100$$

where $\overline{X}$ = mean of the replicate samples
and S = standard deviation

To calculate S, use the formula in #2 on page 192 where:

X_i = the measured value of each replicate
$\overline{X}$ = mean of the measured replicate values
and n = number of replicates

If you feel good about the level of precision that you achieved with your first set of replicates, you can list it as your precision objective. If you feel you could improve your sampling and testing methods to increase your precision, try more replicates until you're satisfied with your results. Make sure, however, that in all future monitoring, you use the same methods used to obtain the level of precision as listed in your objectives.

B. Representativeness

Representativeness refers to how well data accurately and precisely represent actual conditions in your stream or watershed. Your results will be representative if you use consistent data collection procedures, make minimal errors when collecting and analyzing data, and select sites properly. Thus, representativeness goes one step beyond accuracy and precision to address the importance of selecting appropriate sampling sites.

As you know, your monitoring results will vary for different areas along your stream's continuum. In order to describe your objectives for representativeness, you first need to decide what areas of your stream's continuum you want to cover in your monitoring program. Then the next step is to decide how well you hope to evaluate the variability within those areas. The number of sites your Streamkeeper group can monitor and the number of samples they can collect and analyze will be important considerations in making these decisions.

For example, let's say you are evaluating the "bug" community in a stream reach with an even pool to riffle ratio. A simple representativeness objective would be to analyze samples from one pool and one riffle. A more rigorous representativeness objective might lead you to sampling three riffles and three pools.

If you're evaluating bugs in a low gradient (primarily non-riffle) reach, the habitat you'll sample to obtain representative data will be different. If the bug habitat in the reach consists of 70% woody debris, 20% aquatic plants, and 10% leaf litter, you might sample each habitat in proportion to how much they represent the reach. If you know your Streamkeepers can handle 20 samples, taking 14 woody debris, 4 aquatic plant, and 2 leaf litter samples should yield the most representative results.

C. Comparability

Comparability refers to how well different sets of data can be compared to one another. How you design your monitoring program to ensure comparability depends on what you hope to compare. If you wish to compare data collected at different times of the year, you need to make sure you collect the data from the same site each season.

If you wish to compare data from different sites, it's best to collect the data from the different sites simultaneously. If this is not possible, try to schedule data collection in a way that minimizes variation from weather and other factors. For example, flow data would not be comparable if measured on a Monday at one site and a Wednesday at a second site if a significant rainstorm occurred on the Tuesday in between.

As discussed in Chapter 3, if you're comparing different sites as part of an impact assessment, you should select the sites such that natural variation is minimized. Critical habitat characteristics, stream order, topography, gradient, geology, and hydrology all need to be included in your site selection criteria.

No matter what type of comparisons you hope to make, your data collection and data analysis methods need to be consistent in order to achieve comparability. If there's a resource management agency or other entity collecting data on your stream, consider using the same methods and report your results in the same units so your data will be comparable to theirs. State clearly in your QA plan what you hope to be able to compare and how you plan to ensure comparability.

D. Completeness

Completeness is the percentage of your data judged to be valid. To ensure completeness, you'll want to avoid errors and loss of data. Make sure you collect a sufficient number of samples and label them carefully, especially if they're not going to be analyzed right away. Make copies of your data and store it where it won't get lost or misplaced.

To calculate percent completeness, use the following formula:

$$\text{Percent Completeness} = (\%C) =$$

$$(V / T) \times 100$$

where V = the number of measurements judged valid

and T = the total number of measurements

For example, let's say you had ten monitoring teams collecting data on the same day at the same time. If one team had car trouble and couldn't reach their site, and another lost their samples before they could be analyzed, your data set for that day would only be 80% complete. Generally, an 80% to 90% completeness rate is considered acceptable. Decide what you'd like your completeness objectives to be and state them in your QA plan.

I NEED MORE SAMPLES!

Summarizing your DQO's

You may find it helpful to outline your data quality objectives in chart form. An example prepared by the Chesapeake Bay Citizen Monitoring Program appears below. In addition to listing DQO's for accuracy and precision, the chart lists the method used to evaluate each parameter and the range, units, and sensitivity of each method.

The sensitivity of a method refers to how small or large the increments for reading the results are. For example, the Chesapeake Bay Program uses pH kits that can read dissolved oxygen in increments of 0.1 mg/l. This is a more sensitive method than one that can only read it in increments of 0.5 mg/l.

If you wish, you could extend this table to include a column for representativeness, comparability, and completeness. However, representativeness and comparability DQO's may be more easily described in narrative form.

8. Project narrative

The purpose of a project narrative is to make sure that readers of your QA plan can relate your project to the DQOs listed in section I-7 and to the project description outlined in section I-6. A list of elements that should be included in your project narrative follow.

Note: you will find that most of the elements at the top of page 197 are addressed in other sections of your QA plan. You only need to include narratives of those elements that are not described in detail in other sections.

TABLE 14 SAMPLE DATA QUALITY OBJECTIVES

Parameter	Method/Range	Units	Sensitivity	Precision	Accuracy	Calibration
Temperature	Thermometer -5.0° to +45°	ºC	0.5 ºC	±1.0	±0.5	with NBS.Certified Thermometers
pH	color comparator wide-range	Standard pH units	0.5 units	±0.6	±0.4	Orion Field pH meter
	narrow range		0.1	? *	±0.2	Beckman pH meter
Salinity	Hydrometer	parts per thousand (0/00)	0.1 0/00	±1.0	±0.82	Certified Salinity Hydrometer Set
Dissolved Oxygen	Micro Winkler Titration	mg/l	0.1 mg/l	±0.9	±0.3	Standard Winkler & Y.S.I. DO Meter
Limit of Visibility	Secchi Disk Depth	meters	0.05 mg/l	NA	NA	NA

*lack of sufficient data

Project Narrative Elements

- anticipated use(s) of the data
- how the success of the project will be determined
- a description of the survey design
- the types and locations of samples required
- sample handling and custody requirements
- methods for analyzing data
- calibration procedures and use of replicate samples
- standard operating procedures for field sampling
- plans for peer reviews prior to data collection
- ongoing assessments that occur during the course of the project

9. Special Training Requirements/ Certification

When thinking about assigning monitoring tasks to a "Streamkeeper recruit," you must make sure he or she can do the job. Self taught individuals and groups that begin monitoring efforts on their own can be very effective, but individuals and groups that receive training are capable of producing credible data much more rapidly.

In this section, list all training that you or your fellow Streamkeepers will need, and any certificates obtained. For example, Streamkeepers trained by the Adopt-A-Stream Foundation complete a ten-hour training program that includes three hours in the classroom, and seven hours in the field. That training is based on material presented in this book, and our other publications, *Adopting A Stream: A Northwest Handbook*, and *Adopting A Wetland: A Northwest Guide*. Each graduating student receives a Streamkeeper Graduation Certificate and has the opportunity of earning graduate credit from several universities which have accredited the course.

The States of Texas and Kentucky have established outstanding training and certification programs for water quality monitoring. And in British Columbia, the Federal Department of Fisheries and Oceans has developed a Streamkeeper training and certification program in conjunction with Capilano College. Check around, and you will likely find an instruction program near you that fits your training needs.

In this section of your QA plan, clearly define your training requirements. In addition to training for routine inventory and monitoring activities, you may need special training to deal with collection and handling procedures of water quality samples for lab analysis, or for conducting fish and wildlife surveys. Add descriptions of any specialty training requirements to this section.

Finally, be aware that your monitoring results will gain another "credibility notch" if you build in a regular training schedule into your program design.

10. Documentation and Records

Part of your QA Plan is the documentation and records necessary to answer any questions that may arise. You should include the following:

- copies of all data collection forms
- a catalogue of raw data (note storage location and access requirements)
- a roster of all participating Streamkeepers' names, addresses, and telephone numbers
- a roster of all supporting experts
- copies of all important correspondence
- receipts and other financial records

II. DATA ACQUISITION

1. Sampling Design / Sampling Methods Requirements

In section I-6 of your QA plan, "Project Description," you will find a description of how you designed your monitoring study. It should include a list of parameters, the number and types of samples, the required sampling equipment, and the methods for evaluating data. In this section, describe your sampling design in more detail and describe the specific methods it requires. When describing sampling method requirements, include the following items:

- **Standard Operating Procedures** (SOPs): Prepare directions for each procedure. This can be done simply by including copies of instructions from this book, other guides, or instructions from manufacturers of any apparatus that you use. The SOPs should also include directions on how many replicate samples are required. Because they will be used in the field repeatedly, you may want to list the SOPs in an appendix at the back of your QA plan so they can be easily referenced.

Make sure the procedures and equipment you plan to use will actually work for you in the field. Test out your physical and biological monitoring procedures with your fellow Streamkeepers to make sure everyone can do them well. Then, make those procedures your SOPs. Anything that seems too complicated will generally turn out to be frustrating and, perhaps not worth the effort. If you keep it simple you will minimize errors.

• **Guidelines Used to Select Monitoring Sites**: Describe the rationale that you used in selecting your monitoring sites.

• **Habitat Characteristics:** Provide a general description of each monitoring site. For example, "stream site #1: cobble bottom; approximately 2.5 meters wide; 10 centimeter average depth, dense riparian vegetation."

This will help ensure that Streamkeepers make observations only in areas that represent the habitats you are trying to evaluate.

• **Sampling Schedule:** Prepare a sampling schedule for each site or each parameter. Take measures to avoid potential gaps in your data collection process, such as scheduling backup Streamkeepers to monitor when someone is ill or on vacation. If you rely on students to do your monitoring, make sure that the summer schedule is covered.

• **Documentation**: List the types of documentation and reports that will result from your study. Establish a procedure to spot check data collection forms to ensure that they are filled out completely and correctly.

• **QC Requirements:** List the QC steps built into your program, including number of replicates used.

• **Method of Gear Selection:** Describe why you selected the monitoring equipment that you use. For example, you may have selected a particular mesh size on a net because it was considered the optimum size for collecting

the types of bugs found in your stream. You may have selected low cost test kits for economic reasons, knowing that the resulting data may have limited application. Or maybe you're using some high tech equipment because it was recommended by an agency and they loaned it to you free of charge.

- **Level of Effort**: Describe how much time and effort is required for each monitoring activity. To a certain degree, your sampling schedule describes that effort. Additionally, you will want to describe the effort associated with recording data, preparing reports, and presenting your results. Also, describe the numbers of people involved. You may find it makes sense to start out with simple monitoring procedures and move on to more sophisticated methods as you get more experience and training. As your program evolves, amend your QA plan accordingly.

2. Sample Handling and Custody Requirements

Your QA plan should identify who is collecting samples and making observations. For chemical analyses, identify whether or not the analysis is being done in the field, back home, or in a lab.

Detail any special handling requirements. For example, dissolved oxygen samples must be analyzed right away, or "fixed" with the first two reagents in your test kit, because the concentration will change during transport.

Make sure all field samples (chemical or biological) are properly labeled with date, time, site, and the name of the person who collected the sample.

If you are using samples for legal purposes, the **chain of custody** procedure must be used. This procedure creates a written record that can be used to trace the possession of the sample from the moment of collection throughout the entire data analysis. According to the *EPA Volunteer Monitoring Guide: A Guide for State Managers*, the minimum documentation that should be used is described in the following column:

Field Sampling

- outline procedures for preparing reagents or supplies that become an integral part of the sample
- outline procedures for locating sampling sites and other considerations associated with sample acquisition; use standard forms for locating, describing, and recording sampling locations
- document sample preservation methods
- use pre-prepared sample labels containing all information necessary for effective sample tracking
- use standardized field tracking reporting forms to establish sample custody in the field prior to transport

Laboratory Analysis

- identify who will be sample custodian at the lab facility and who will be authorized to sign for incoming field samples
- establish a lab sample custody log consisting of serially numbered, standard lab tracking report sheets
- outline lab sample custody procedures for handling, storing, and distributing samples for analysis

3. Analytical Methods Requirements

Section II-1 provides Standard Operating Procedures for collecting samples. In this section you will prepare guidelines for analyzing the samples collected.

For example, macroinvertebrate collection procedures you prepare should include directions on equipment used (net type/mesh size), and collection techniques (rock rubbing/kicking substrate). In this section you will prepare directions on how samples will be analyzed, which includes the level of identification (order, family, etc.), and the types of metrics and statistical analyses used. Develop descriptions of your analytical methods for flow, water quality, and other parameters as well. Site any literature that you use for identifying and analyzing samples.

4. Quality Control Requirements

Prepare a checklist of quality control (QC) procedures for use with equipment maintenance, field activities, and data analysis. Some elements of a typical QC checklist are listed below.

Equipment Quality Control

- check to make sure equipment is in working order and not damaged
- clean equipment before and after taking it into the field
- label equipment with their dates of purchase and dates of last usage
- check the expiration date of chemical reagents prior to each use
- check the batteries of all equipment that requires them
- make sure equipment is calibrated appropriately before conducting each test

Field Procedures Quality Control

- collect replicate samples
- conduct repeat and/or side-by-side tests performed by separate field crews
- occasionally mix and alternate individuals within field crews to maintain objectivity and minimize individual bias
- review field records before submitting for analysis to minimize errors

Data Analysis Quality Control

- check all calculations twice
- hard copies of all computer entered data should be reviewed for errors by comparing to field data sheets
- have qualified professionals review your data analysis methods and results periodically

As you see, a QC checklist is a tool designed to keep your credibility high. Without it, you may get erroneous results because you forget to check the batteries in your stop watch or that fancy electronic meter you acquired. Or you may forget to repeat certain tests. As you gain more experience and your monitoring program grows over time, your QC checklist will also grow.

5. Instrument / Equipment Testing, Inspection Requirements

This section of your QA plan is very simple. Prepare a chart that lists the inventory and monitoring equipment that you will be using for your study. Next to that list, describe all testing, inspection and maintenance requirements for each apparatus. Note that "consumables" such as disposable parts of test kits and reagents, are treated below in section I-7, "Inspection/Acceptance Requirements for Supplies and Consumables."

You should also include a brief narrative describing where and how your inventory and monitoring equipment is stored and if you have any security requirements/procedures.

6. Instrument Calibration and Frequency

In this section, describe in detail your procedures for calibrating your equipment. Information about how to calibrate equipment used to measure various parameters can be found on pages 192-193 in section I-7, "Data Quality Objectives." You should also set up a log so you can ensure and demonstrate that your equipment is being calibrated regularly.

7. Inspection / Acceptance Requirements for Supplies and Consumables

Prepare a list of inventory and monitoring supplies and consumables. These items are not included on your inventory list from section II-5 because they are not considered permanent pieces of equipment. Keep a record of the items on hand, when they were purchased, and dates when different materials should be replaced. This inventory will assist you in keeping an adequate stock of supplies and consumables on hand. It may also help you when you are preparing requests for financial assistance.

In addition to preparing an inventory, describe how and by whom supplies and consumables will be inspected and accepted for use in the project. You should also establish security procedures for storage of your consumables. One Streamkeeper reported that some of her packets of a powder reagent disappeared one day, and were later discovered in the pockets of a middle school student expecting to "get high." Your QA plan should reflect any security measures that you employ to avoid the loss of consumables, and more importantly, the safety of participating Streamkeepers.

8. Data Acquisition Requirements for Non-direct Measurements

Remember, data is "any information used to make a decision." It appears in many forms, not just as a list of numbers or bugs. Some of the data you use to design your study and analyze your results may be information you did not measure directly, but instead gathered from other sources.

In this section, prepare a summary of the types of "non-direct measurement" data you used in your project and document their sources. For example, if you compared land use information with your water quality results, you should record the sources of that information, be it aerial photographs, city zoning and comprehensive plan maps, U.S.G.S. topographic maps, or another source.

To assure the quality of this type of data, you'll need to define criteria that you'll use to accept and reject it for use in your project. Identify limitations on the use of the data and explain the nature of its uncertainty. Keep in mind that even with limitations, some forms of data can be a valuable addition to your study.

Anecdotal information gathered by talking to "old timers" can help you piece together the history of many of our streams. Although it may lack "accuracy" and "precision," this type of information can be helpful especially for streams whose history has not been formally recorded.

9. Data Management

This section is very simple but essential. You should record:

- how your information is being stored (on paper, in a computer, in a file cabinet, etc.)
- where the data is stored
- how the information is accessed for information and data entry purposes (note security measures such as computer passwords or lock combinations)

III. ASSESSMENT/OVERSIGHT

1. Assessments and Response Actions

In this section, you will identify how you plan to assess the capability and performance of your project. Assessments should be done both internally, by members of your Streamkeeper group, and externally, by a qualified outside expert.

Assessments include **system audits** and **performance audits**. A system audit evaluates the process of the overall project, including on-site reviews of field sites and facilities where data is processed and analyzed. A performance audit evaluates how well the people who are in the business of collecting and analyzing data do their jobs.

The purpose of a performance evaluation is to provide positive reinforcement for the actions that are being accomplished well, and to improve the rough areas. A successful evaluation will result in confident Streamkeepers collecting credible information.

Your QA plan should list all parties who assist in these evaluations, their relationship to the project, and how often the audits occur. We recommend you bring in an outside expert at least once a year. With a little investigation, good telephone skills, and perhaps the price of a dinner or lunch, you should be able to find local experts, such as agency biologists or health district water quality specialists, who are willing to help.

Your QA plan should also describe what corrective actions you will take in response to any shortfalls that become apparent from your audits. **Immediate corrective actions** are usually in response to internal audits or QC checks. They may include things like correcting a calibration or sampling procedure. **Long term corrective actions** are usually in response to external audits and may include providing Streamkeepers with "refresher" training on technical aspects of your monitoring program or rescheduling field activities to ensure that a proper sampling design is achieved.

2. Reports to Management

This is another simple but essential QA element. Create a written record of each field exercise, listing who collected the data, where it was collected, the type of study completed, and the date the monitoring activity was accomplished. This way the project manager has a formal report of all activities that took place to implement the project

For large monitoring efforts involving many participants, one Streamkeeper from each site or field exercise should be assigned the responsibility of providing the project manager with the necessary information to complete this report.

IV. DATA VALIDATION AND USABILITY

1. Data Review, Validation and Verification Requirements

In order to ensure that your data was properly collected and recorded, you need to set up a review, validation and verification process. The purpose of this section is to establish the criteria that you will use when reviewing and validating the data you collect.

Validating refers to either accepting, rejecting, or qualifying data before it is used for analysis and/or decision making. This often involves rejecting "outlyers," or individual datum values that seem to be far from accurate. For example, of three replicate dissolved oxygen values reading 9.0, 9.5, and 5.0 mg/l, you may decide to reject the 5.0 reading because it varies significantly from the others.

2. Validation and Verification Methods

This section outlines the specific procedures to be used for reviewing, validating, and verifying your data, based on the criteria established in section IV-1.

As a first step, make sure your data collection forms are standardized. Standardized forms can be more easily spot-checked to ensure they were filled out correctly and completely.

Next, have certain members of your group be responsible for reviewing the recorded data before it is stored in a computer or file cabinet. If the data is put into a computer, guard against data entry errors by establishing a regular schedule to compare raw data with computerized data.

Finally, describe how each of your monitoring results will be confirmed. For example, you may consult an entomologist to verify the identity of a macroinvertebrate, then preserve a sample in alcohol for comparison with all future samples. Visual comparisons with the preserved sample, which is called a **voucher** sample, are then all you need to validate future samples. Make sure you record the procedures that the entomologist used to verify the sample, e.g., ex-

amination under a stereo microscope and consultation with a dichotomous key from a particular entomology text.

It is important to include both internal and external review, validation, and verification of your data, i.e., have both members of your Streamkeeper group and outside experts participate in the process.

3. Reconciliation with Data Quality Objectives

Finally, the last part!!!

In section I-7 of your QA plan you outlined your Data Quality Objectives (DQO's). You established goals for accuracy, precision, representativeness, comparability, and completeness of your data. The purpose of this section is to describe how your project results will be reconciled with your DQO's.

In other words, describe how and when you will determine if your data meets your objectives, and what you will do if it does not.

It is best to assess your data as soon as possible after it is collected so you can begin corrective actions sooner rather than later. If you find limitations in your data, they need to be identified and reported to the project manager and data users.

If you elect to use all of the inventory and monitoring procedures outlined in this book, you will have a broad view of your stream's physical, biological, and chemical characteristics. If you develop a QA plan to ensure that your data is accurate, precise, representative, comparable, and complete, then you will be in a powerful position. You'll be able to use your data to effect positive change in your watershed.

## CHAPTER NINE

Presenting Your Data

A Data Reminder

Remember, the Oxford American Dictionary defines data as "facts or information to be used as a basis for discussing or deciding something." Data is not limited to physical, biological and chemical information. Social and political information is also data.

By now you have conducted a watershed inventory. You have compiled a library of information about your stream and its surrounding watershed. You know its history. You know who owns the land in your watershed.

Friends, neighbors, schools, and community group members have taken an interest in the stream. Collectively you have put together a great stream monitoring quality assurance plan which will serve as a blueprint for all new Streamkeepers in your watershed to follow in the future.

You have identified monitoring stations, and determined which monitoring activities are required to get the information you need. Regular monitoring dates have been established. And you and your fellow Streamkeepers are periodically checking the field procedures being used, ensuring that all monitoring equipment is functioning.

You are getting great information! You have sampled water quality, measured stream flow, and examined the stream bottom. You have a photographic record of your stream as it flows through the seasons. You have photo documented all problems that you have discovered.

You also have a pretty good understanding of the "critters" in your watershed, including the humans. You are looking beyond the "babbling brook" and checking on land use changes in the surrounding watershed. You have located all kinds of water resource problems that need fixing . . . your file cabinets and computers are just brimming over with DATA.

So, what are you going to do with all this great information? By itself, raw data can be informative and, as the principal collector(s), you and your fellow Streamkeepers can be proud of your accomplishments. However, most of your audiences will need your Streamkeeper data presented in a form they can understand. Raw data can be converted to tables, charts and graphs that have a real impact on a wide range of audiences.

TABLES, CHARTS, AND GRAPHS

A good first step for presenting your data is to construct a **table** summarizing results obtained over a period of time. Such a table might include the mean, minimum, and maximum values for each parameter measured at each site. A summary table will make it easier for you to detect trends and to compare your results with the water quality standards that have been established for your stream.

figure 9.1

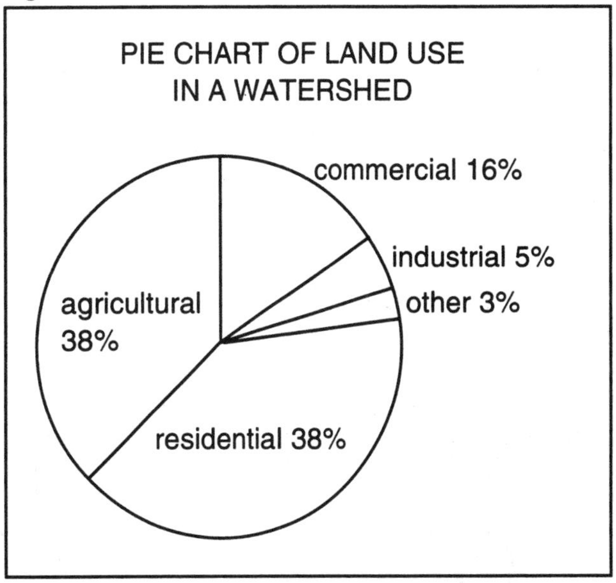

PIE CHART OF LAND USE IN A WATERSHED

commercial 16%
industrial 5%
other 3%
agricultural 38%
residential 38%

TABLE 15 SUMMARIZING ONE YEAR OF STREAM DATA					
parameter	annual mean	minimum	(date)	maximum	(date)
pH	7.1	6.3	(10/15)	8.3	(4/4)
dissolved oxygen	8.5	6.5	(9/8)	13.2	(5/19)
temperature	14°C	5°C	(1/14)	21°C	(8/15)
flow	9.8 CFS	4.1 CFS	(8/15)	19.4 CFS	(3/28)
EPT richness	11	6	(1/14)	15	(9/28)

Once you summarize your data in tables, the next step is to make it even more meaningful to the eye by presenting it in a graphical form. Some basic types of graphs that Streamkeepers can use to present their data are described below.

A **pie chart**, like figure 9.1 at the top of this page, compares parts to the whole. In a pie chart, each value in the data set is represented by a "piece of the pie," with the pie equaling 100 percent of the total values in the data set.

Line graphs and **bar graphs** show how one parameter changes in relation to another. For example, the line graph on the following page, produced by students from Evergreen Middle School in Seattle, Washington, shows how the width of Thornton Creek changes over time.

figure 9.2

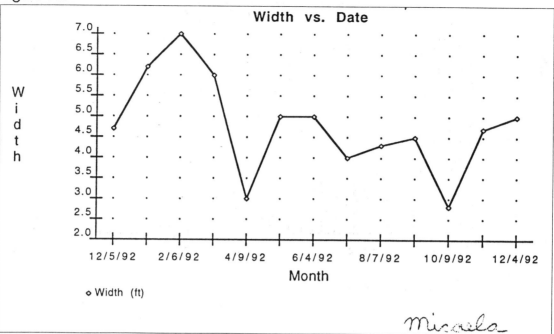

In figure 9.2 above, time is the **independent variable** because it is a constant, and not affected by other factors. Width is the **dependent variable** because it fluctuates in response to other factors, such as volume of water. In bar and line graphs, the vertical "y" axis usually represents the dependent variable while the horizontal "x" axis represents the independent variable.

A line graph, like figure 9.2 shown above, depicts **continuous change** that occurs in a dependent variable. This line graph shows how the width of Thorton Creek changes continuously over time. In contrast, a bar graph uses column bars of varying lengths to depict **individual values**. The bar graph below, also produced by Evergreen Middle School students, emphasizes individual temperature values measured in Thorton Creek on specific dates.

There are endless ways you can use these basic types of graphs to present your data. The next part of this chapter will provide you with some starting points relevant to the types of information introduced in this book.

figure 9.3

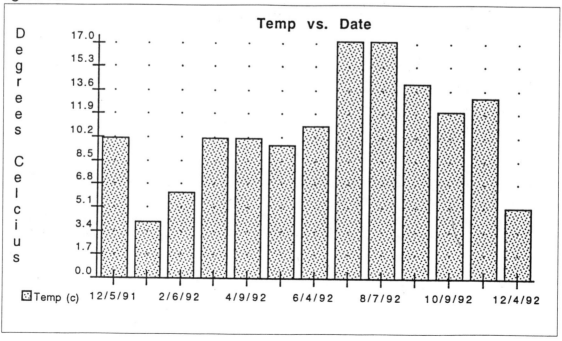

figure 9.4

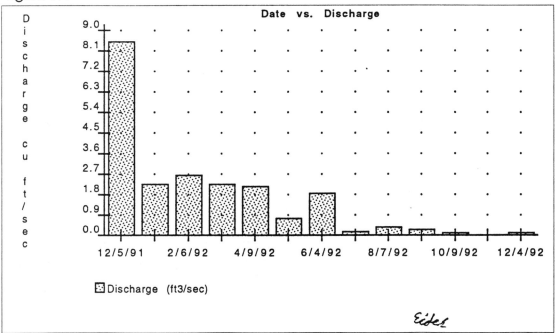

FLOW DATA

In Chapter 5, you learned how to record the flow or discharge of your stream over time. You can use a bar graph to display the flow values that you record each month on your stream. Figures 9.4 and 9.5, provided courtesy of Evergreen Middle School students, depicts monthly flow values of Thorton Creek over the course of one year.

Note the great difference between the flow value measured in December 1991 and one year later in December 1992. No doubt, the first week of December in 1991 was considerably wetter than the first week of December in 1992. This comparison underscores the importance of considering precipitation patterns when you analyze your flow data.

If you connect the tops of each bar in the graph above with a line (figure 9.5), you create a **hydrograph.** A hydrograph, or "water" graph, is a line graph depicting the flow regime of a stream.

figure 9.5

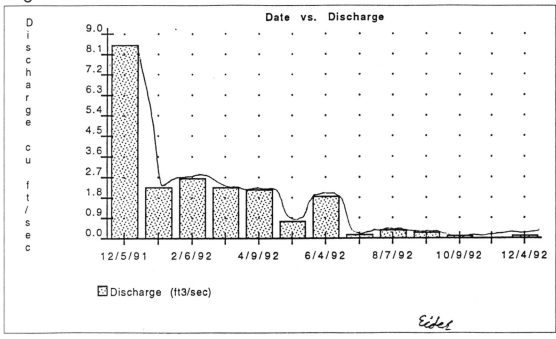

You can use a bar graph to illustrate precipitation as well as flow. Plot the amount of precipitation on the vertical axis with time (days, months, etc.) on the horizontal axis. By connecting the tops of the bar graph, you create another line graph. If you want to get fancy, you can call it a "precipitation index."

You can also present your hydrograph and precipitation index together to graphically illustrate the relationship between the amount of rainfall and the flow of water in your stream. The storm hydrograph on page 106 in Chapter 5 illustrates this relationship for the time just before and after a storm.

Another relationship you may want to illustrate is the connection between an increase in flow fluctuations in your stream and an increase in impervious surfaces in your watershed. Data that illustrates this connection can help support efforts to protect wetlands or riparian corridors from development.

For example, figure 9.6 below illustrates how wetland loss can affect the hydrology of a stream. It compares two storm hydrographs, one before wetlands were removed, and one after they were lost to development. Perhaps you have measured the flow of your stream during rainstorms and seen this same trend occur in your data as development increased in your watershed. You can overlay "before" and "after" storm hydrographs in a similar manner to illustrate the effect of

figure 9.6

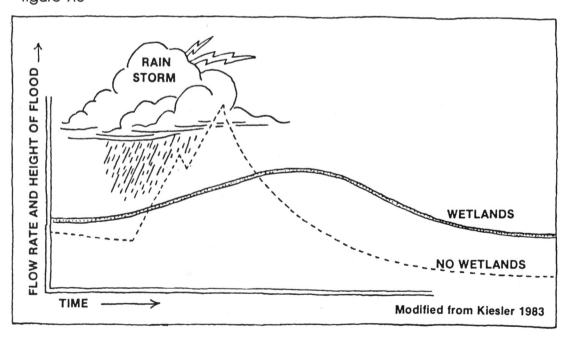

Reprinted from *A Field Guide to Kentucky Lakes and Wetlands,* courtesy Kentucky Natural Resources and Environmental Protection Cabinet.

The Streamkeeper's Field Guide

development on your stream's hydrologic integrity.

By adding a cartoon to your presentation, like the one on the left, you can add a lifelike "punch" to your data. A car driving in "wheel deep" water helps members of your audience realize the implications of your data and think more about how precipitation, flow, and impervious surfaces affect their lives.

Remember that too much water is not the only problem associated with stream flow. In areas prone to drought, water availability can be a critical issue. Often the aquatic critters and other wildlife that rely on streams and riparian areas get left "high and dry" when water is diverted for agriculture and power production. The bar grap in figure 9.7 from the *Yakima Basin Resource News* clearly depicts the critical water situation that occurred in eastern Washington State in 1993.

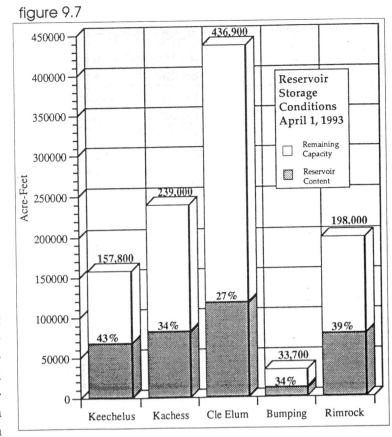
figure 9.7

figure 9-8

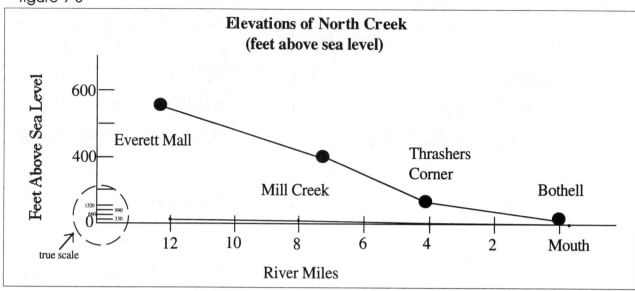

Elevations of North Creek
(feet above sea level)

GRADIENT DATA

You can illustrate the gradient of your stream using a simple line graph like figure 9.8 above. Plot elevation on the vertical axis and river mile or some other distance measurement on the horizontal axis.

Note: Be careful how you present your data. In this chart, distance is depicted differently on the "x" axis than on the "y" axis, making it seem that the gradient of North Creek is very steep, when in fact, the stream has a low gradient, a mere 600 foot change over twelve

miles. (Recommended reading: *How to Lie with Statistics*, by Darrell Huff, W.W. Norton and Co. This small paperback, now in its 43rd printing, will help prepare you to look carefully and ask questions about data that may be presented by your opposition.)

WATER QUALITY DATA

Most water quality data can be presented with line or bar graphs. A simple line graph can depict how the water quality at a particular site changes over time. You can also overlay data from more

figure 9.9

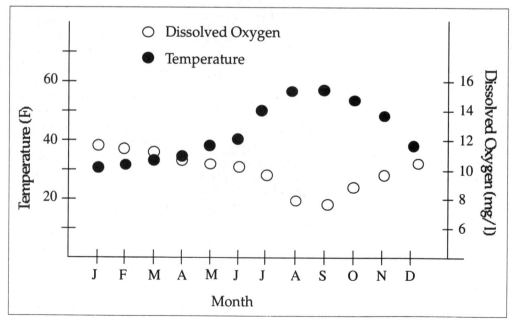

figure 9.10

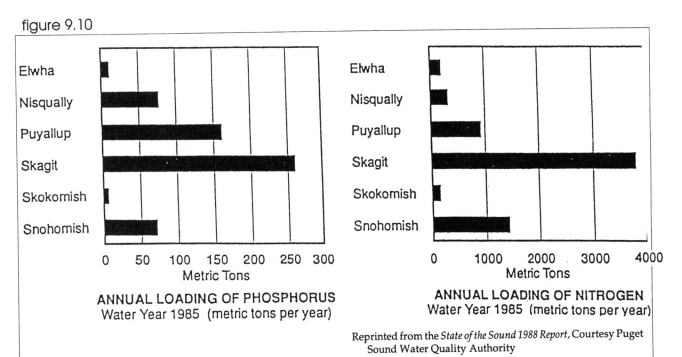

ANNUAL LOADING OF PHOSPHORUS
Water Year 1985 (metric tons per year)

ANNUAL LOADING OF NITROGEN
Water Year 1985 (metric tons per year)

Reprinted from the *State of the Sound 1988 Report*, Courtesy Puget Sound Water Quality Authority

than one parameter on the same plot. For example, the plot shown in figure 9.11 illustrates a close relationship between rising temperature and decreasing dissolved oxygen.

You may also want to present water quality data from different sites on the same graph. Bar graphs like figure 9-10 above are better for this than line graphs because the variable "site" represents distinct, rather than continuous, values. These bar graphs illustrate the annual loading of phosphorous and nitrogen in some of the major river systems in the Puget Sound Basin in Washington State.

You can present data from different sites on your stream in a similar manner. If you present data from different sites on the same graph, make sure all the data for each site was collected over the same time period. You cannot make a meaningful comparison between different sites with a graph that presents data that was collected over inconsistent time periods.

Pie graphs are also useful for presenting water quality data. The Bay Nutrient graph (figure 9.11) uses pie graphs effectively, breaking out nitrogen and phosphorus loadings into Chesapeake Bay by source. The consequences of these nutrient loadings are then described in the space adjacent to the graphs.

figure 9.11

Bay Nutrient Pollution Sources

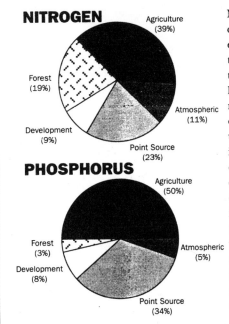

Nitrogen and phosphorus are called nutrients because they are essential to plant growth. When they are available in low amounts, they cause no problems. Excessive levels, however, accelerate the natural "aging" process of bodies of water: they cause the growth of undesirable algae, thereby reducing water clarity, and under some conditions, depleting the oxygen essential to aquatic life. Adequate dissolved oxygen is critical for finfish, particularly in their earliest stages. Oysters, clams, and worms which cannot move to any oxygen-rich area, will die when levels are inadequate. *Point sources of pollution are pipes. Nonpoint sources are runoff or atmospheric deposition.*

Courtesy the Interstate Commission on the Potomac River Basin

COOL TEMPERATURE 90% OF SATURATION

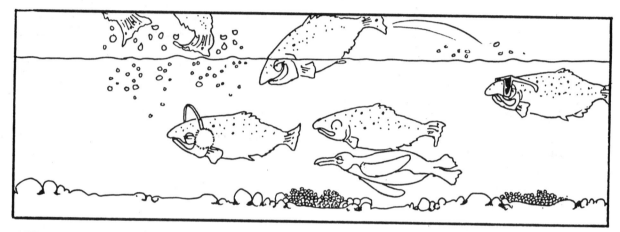

Water quality data have great potential to be dry. Add life to your presentations whenever possible. For example, try showing a school of happy fish at the locations in your stream where there are high saturation levels of oxygen, and a few unhappy or deceased fish at locations where oxygen levels are very low. This type of graphic will not only liven up your presentations, but will bring home the real meaning of your findings to any audience.

MACROINVERTEBRATE DATA

This book introduces you to using macroinvertebrates as biological indicators of a stream's health. There are a variety of ways to present data on these creatures.

To get a better understanding of your stream's macroinvertebrate community, use data presentations to help you analyze how the community changes over time and varies under different conditions. There are endless comparisons you could make. For starters, you may want to

plot abundance (density), taxa richness, and EPT richness over time, to see how these factors change during the course of a year. A multi-tiered bar chart (see figure 9.13) could work for this analysis.

figure 9.13

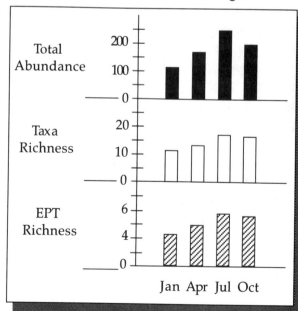

The Streamkeeper's Field Guide

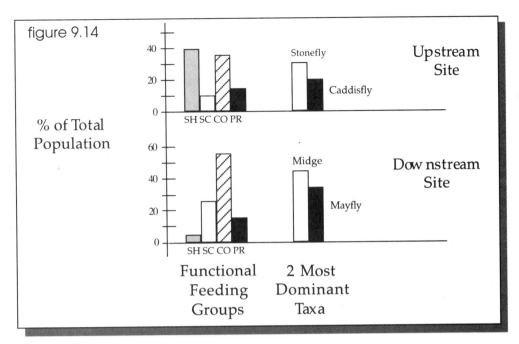

figure 9.14

% of Total Population

Upstream Site

Stonefly

Caddisfly

SH SC CO PR

Downstream Site

Midge

Mayfly

SH SC CO PR

Functional Feeding Groups

2 Most Dominant Taxa

You may also want to compare these factors for different study site locations, and/or habitat types. In this case, you could use the same type of histogram chart, replacing the dates on the "x" axis with specific study site locations or different habitat types, such as pool versus riffle or high versus low gradient.

A multi-tiered bar chart will also work well for presenting information on functional feeding groups and dominance. Figure 9.14 compares these two factors for two different sites.

You could use this format to compare other types of habitat differences or to compare populations in different times of the year.

If you have used a pristine reference stream to compare to your stream, and found that your stream is not as healthy as it should be because of a suspected impact, you could use a chart similar to figure 9.15 to present the comparison.

When using a reference stream, make sure that variability in natural factors are controlled as much as possible. Be prepared to provide evidence that your stream is similar to the reference stream ecologically if the impact was not present.

This type of graph can also be used to show the effect of an impact on your stream if you compare a site upstream from the impact with a downstream site, or if you compare one site before and after the impact event occurred. The shaded bar would represent the upstream or the "before" situation; the unshaded bar would represent the downstream or the "after" situation.

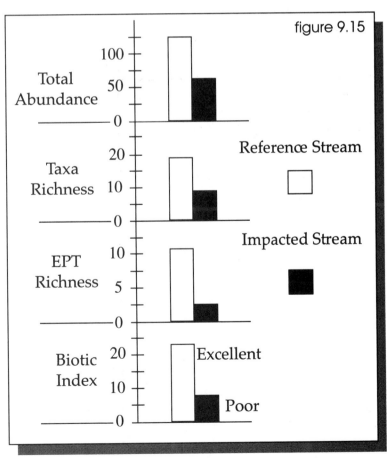

figure 9.15

Total Abundance

Taxa Richness

EPT Richness

Biotic Index

Reference Stream

Impacted Stream

Excellent

Poor

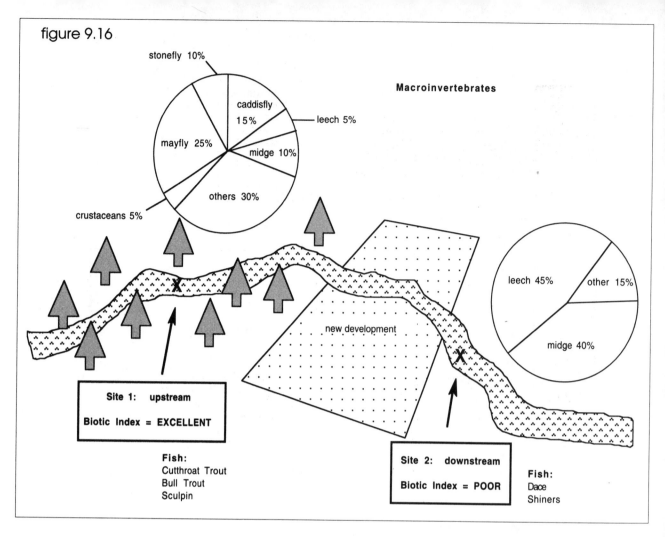

figure 9.16

stonefly 10%
caddisfly 15%
leech 5%
mayfly 25%
midge 10%
others 30%
crustaceans 5%

Macroinvertebrates

leech 45%
other 15%
midge 40%

new development

Site 1: upstream

Biotic Index = EXCELLENT

Fish:
Cutthroat Trout
Bull Trout
Sculpin

Site 2: downstream

Biotic Index = POOR

Fish:
Dace
Shiners

Pie charts are useful for presenting information on the relative abundance of different organisms, to show dominance patterns, proportions of functional feeding groups, or overall diversity. Figure 9.16 above uses pie charts to compare the relative abundance of different organisms at two sites.

You can add even more life to your presentations by using sketches of the macroinvertebrates themselves, as well as pictures of the fish which depend on them for food.

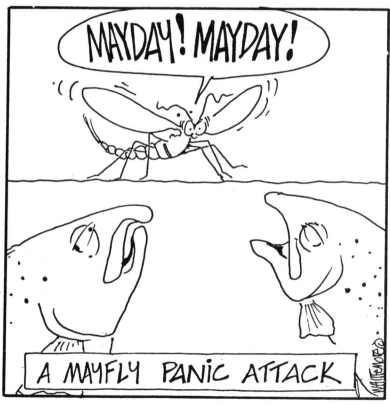

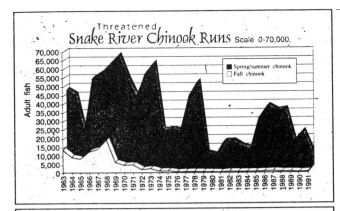

Threatened Snake River Chinook Runs Scale 0-70,000.

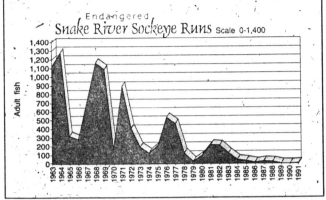

Endangered Snake River Sockeye Runs Scale 0-1,400

figure 9.17

The Status of Snake River Salmon

In 1991 and 1992, Snake River runs of chinook and sockeye salmon were listed under the federal Endangered Species Act. These charts are based on fish counts at dams. Note that the sockeye run is on a different scale, given its extremely low numbers.

Reprinted from *Strategy for Salmon*, courtesy the Northwest Power Planning Council

THE TRAGEDY OF MEMORY LOSS IN WILD SALMON

FISH AND WILDLIFE DATA

Observations of the presence or absence of fish by species can be presented to make a variety of points. After you have been collecting data for a long time your presentation can be very dramatic, like the one above depicting the status of Snake River salmon, figure 9.17.

Figure 9.18, prepared by the Northwest Power Planning Council in *Strategy for Salmon*, provides a historic perspective of the decline of salmon in the Columbia River. Unfortunately, many rivers and streams in the Pacific Northwest have similar histories.

Once again, the more visually interesting you can make your data presentations, the more effective they will be. Figure 9.19 prepared by the Chesapeake Bay Program provides an example of an effective way to present data on wildlife. Adding a duck to a graph of waterfowl counts brings the message of the data more vividly to the observer's eye.

figure 9.18

Columbia River Basin
Salmon Runs An Historical Perspective

figure 9.19

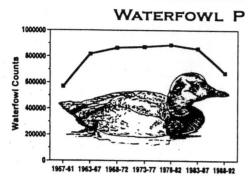

WATERFOWL POPULATIONS ARE VARIABLE

- The Chesapeake Bay is the wintering, breeding, or stopover area for 29 species of waterfowl.
- Today, approximately one million swans, ducks, and geese winter in the Chesapeake Bay region, representing 35% of all waterfowl in the Atlantic Flyway.
- These waterfowl populations have responded differently to man's increasing growth. Mute swan and mallard populations, for instance, have increased because they have adapted to human activities. Black ducks and diving ducks, on the other hand, are more sensitive to disturbances in their habitat, resulting in declining numbers. Overall, waterfowl populations in the Bay region have declined 10-20% in recent years.
- In December 1990, the Chesapeake Executive Council adopted the Chesapeake Bay Waterfowl Policy and Management Plan which will guide the implementation of programs to protect and enhance waterfowl populations and their habitats.

TYING THE PHYSICAL, BIOLOGICAL, AND CHEMICAL DATA TOGETHER

The *Streamkeeper's Field Guide* emphasizes a holistic approach to looking at a stream. You have learned how to consider physical, biological, and chemical aspects of your stream and its surrounding watershed. All things are connected. This gets to the root of the concept of **ecology**, the study of the interrelationships between living species and how biotic and abiotic factors interact.

There are endless ways to pull together your data and graphics to illustrate the ecological interrelationships within your stream and watershed. The example below combines information on algae, macroinvertebrates, fish, water appearance, dissolved oxygen, and the relative location of a particular point source along the river. This type of presentation is a handy tool that can be used in arguments for protecting or enhancing any of the individual elements about which you present information.

A similar type of graphical presentation at the right illustrates the relationship between land use, erosion, embeddedness, and salmonid spawning success. While the same information could be relayed using bar or line graphs, the visual appeal of this presentation makes it more effective.

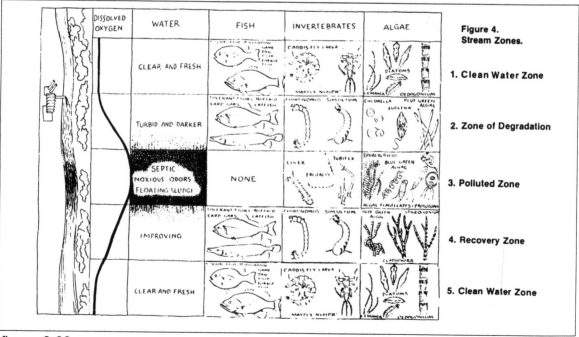

figure 9-20

Reprinted from *A Field Guide to Kentucky Rivers and Streams*, courtesy Kentucky Natural Resources and Environmental Protection Cabinet

figure 9.21

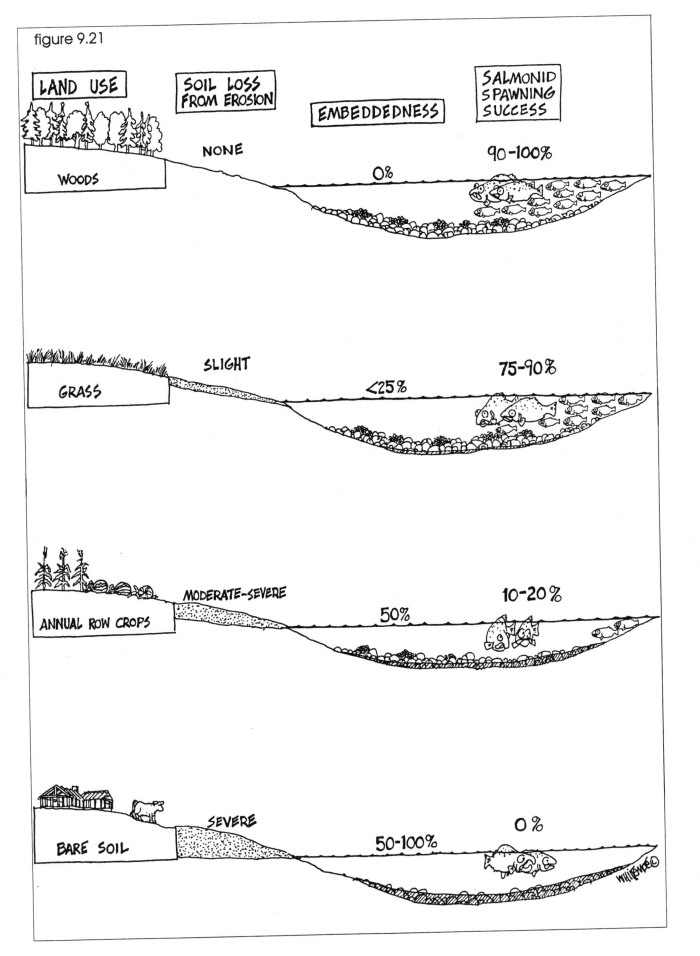

You may have encountered cattle wandering in and around your stream as it flows through a feed lot on a dairy farm. Rob Taylor, reporter from the Seattle, Washington *Post Intelligencer*, researched this type of situation and reported that 200 cattle can produce animal wastes equivalent to the sewage from a town with a population of 4,360.

You could plot out the difference between upstream and downstream reaches in terms of fish populations, macro invertebrate populations, dissolved oxygen levels, and, perhaps, fecal coliform counts. Or, depending on your audience, you may want to depart from the standard methods of presenting data and draw the picture of a contented cow on a toilet with an informative caption, as shown in figure 9.22.

figure 9.22

200 cows can produce the same amount of waste as 4360 people (Rob Taylor, in the Seattle *Post Intelligencer*)

figure 9.23

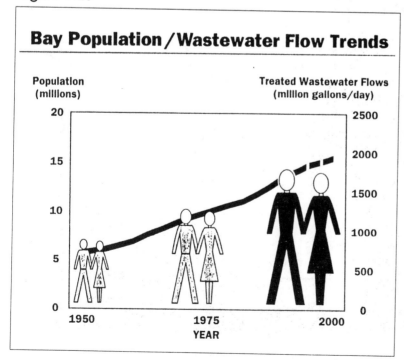

SOCIAL AND POLITICAL DATA

By now you know that being a successful Streamkeeper requires you to be just as aware of the social and political nature of your watershed as the physical, biological, and chemical characteristics of your stream. These factors are intimately intertwined.

During the course of your watershed inventory, you should note the changes in demographics in your watershed. With your other supporting data, you may be able to draw some conclusions that can be presented in a dramatic fashion. For example, figure 9.23 from the Interstate Commission on the Potomac River Basin, shows a correlation between increased population and pollution in the Chesapeake Bay area.

figure 9.24

If forestry activities occur in your watershed, don't focus exclusively on physical and biological observations. There may be some more "telling" bits of information to look at, such as land sales or timber harvest permit applications. Figure 9.24 illustrates the increase in timber harvest applications to Washington State's Department of Natural Resources since 1988.

The Department of Natural Resources attributes the sharp increase between 1992 and 1993 to high timber prices, salvage of blow-down from winter windstorms, a fear of new regulations and a possible ban on exporting logs. Landowners across the state felt they had better cash in their timber before any new regulations or a ban made it difficult to cut and sell.

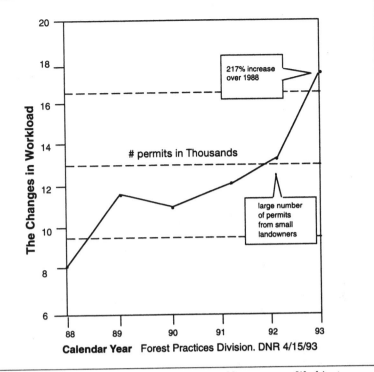

Reprinted from *Forest Resources News*, courtesy Washington Environmental Council Forest Resources Committee

THE TRAGIC CONSEQUENCES OF SIMULTANEOUS CUTBACKS IN WETLANDS AND CROSSING GUARDS.

WHITEMORE

recommend you consider the finny, furry, and feathery critters as user groups also. In your presentations, you can simply provide a list of user groups for your audience, or you might consider taking a more creative tack. In the cartoon below, Brian Basset tells a poignant story about the Snake River sockeye salmon and different user groups on the Columbia River. Indeed, they have an upstream swim ahead.

If you have monitored the attitudes of the elected officials who preside over your watershed, you may want to present an overview of their environmental voting records. Pick a particularly tough issue like wetland protection and you may find that the collective political attitude is one of avoidance. You can present actual voting records of elected officials along with a picture like the one at the top of the next page that tells the "whole story."

It's obvious that current water resource user groups will have to give up something in order for fish species to survive. When you identify the user groups in your watershed, we

BRIAN BASSET ©1991 THE SEATTLE TIMES

ALUMINUM INDUSTRY

IRRIGATION-DEPENDENT FARMERS

ELECTRICAL USERS

DOWN RIVER COMMERCIAL FISHERMAN

SNAKE RIVER SOCKEYE

UTILITIES

UP RIVER BARGE OPERATORS

SALMON RECOVERY PLAN

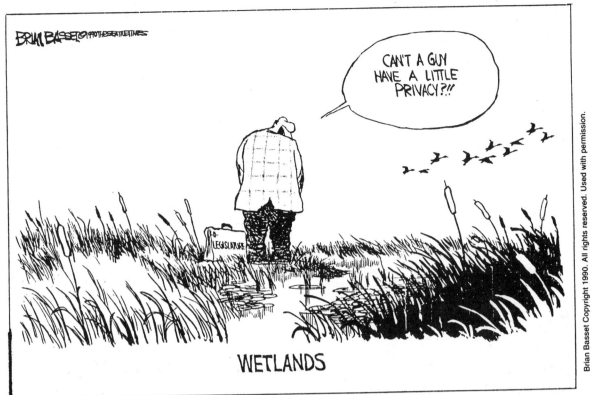

Perhaps you'd like to get a pulse on the environmental attitudes within your watershed. If you do, remember to ask objective, unbiased questions and get a truly random sample of respondents if you want your data to have statistical significance. While the presentation of attitudes reflected in the cartoon below is not statistically valid, the cartoon could be an effective complement to a data presentation on public opinion of the Endangered Species Act. This type of addition to a data presentation may help get your point across.

OTHER DATA PRESENTATION TOOLS

In addition to tables, graphs, charts, figures, and cartoons, there are other handy tools for data presentations. A few are discussed below.

Maps

Maps can be especially useful for presentations about the effect of human development on the natural features of your watershed. While local governments usually review the environmental impacts of new residential, commercial, and industrial developments, they often review each project independently. The cumulative effects of all developments that occur in the watershed are rarely considered.

You can help shed light on the cumulative effect of development in your watershed by producing a watershed map depicting developments that have occurred over a certain period of time. It is helpful to color code different types of development on the map and add a key that lists them and breaks them down by percentage.

You can also keep the information updated on a regular basis. Use plastic overlays on top of a base map of your watershed, with each overlay representing development that occurred within a specific time frame. As the overlays pile up, you can really see the cumulative development that has occurred, and it will become apparent that governments should not review projects independently of each other if they want to maintain the integrity of their watersheds.

The Streamkeeper's Field Guide

Photographs

A good picture adds credibility to any bit of quantitative data. During your investigation phase, you no doubt ran across some historical photographs of your watershed from a variety of sources. If you can, obtain or reproduce copies of these photographs so you can use them in presentations with current photos you have taken.

Comparing "before" and "after" photographs of your watershed can make a strong argument for protecting what is left and restoring what has been lost. Use 35 mm slides and you'll have an inexpensive but effective and professional tool for presenting your stream's case before a large audience. Video tape images can also be very effective if properly edited.

North Creek (WA) watershed 1955

Both photographs courtesy Snohomish County (WA) Planning Department.

High Tech Tools

In this day and age, there is a high probability that you or someone you know has a personal computer. A good software package with a database function will make your job of organizing and presenting data a much easier task. Once you enter your data into a computer database, you can produce the types of line graphs, bar charts, and pie charts discussed in this chapter with only a few taps on your keyboard. You can also obtain software packages to perform statistical analyses on your data.

Computer technology can also really make your data "sing" by integrating graphics, photographic images, and sound. For example, the Washington State University College of Education has put together a *Hypercard* program for Macintosh computers that will display a picture of a watershed and its stream. You can move a cursor "downstream" to a stream inventory site, click the mouse, and a photograph of the site will appear

North Creek watershed 1991

on the screen. The inventory and monitoring data collected at that site will appear on another screen. And all of this data can be projected on a large screen for big audiences.

In order to have the greatest impact, give some real thought to how you want to package all of your data. Raw data, tables, graphs, charts, maps, photographs, and interview information must be surrounded with written word explanations that will make a positive impression on your audience.

Take a few minutes to read the following article by Eleanor Ely, the editor of *The Volunteer Monitor*. By following these tips, you will be able to avoid many common pitfalls.

Tips for Producing Publications That Impress

by Eleanor Ely

As editor of *The Volunteer Monitor*, I probably receive more publications produced by citizen monitoring programs than anyone else in the entire world. I've noticed how a professional-looking, well-written newsletter or brochure gives me an overall feeling of confidence in the program that produced it. Conversely, a sloppily produced, typo-filled publication makes me wonder whether that program is equally careless about data collection. Following are some guidelines for producing publications that make a good impression.

Planning

The guidelines for planning a credible publication are remarkably similar to those for setting up a credible citizen monitoring program. Set clear goals. Know what audience you want to reach. Provide accurate and useful information. Match the presentation to the audience.

The goals and the intended audience for a brochure for the general public will be very different from those for, say, a newsletter for volunteers. The brochure needs to "hook" readers with an eye-catching layout (it may be worthwhile to hire a professional designer) and a clear message explaining why your program should matter to them. The volunteer newsletter, on the other hand, doesn't need to be flashy. But do give the volunteers real news about the program—especially about how their data are being used.

Proofread, proofread, proofread

Careless mistakes can make the best designed publication look amateurish. Never skip that one final proofreading before you go to print; you're sure to find something like the following (from a recently published volunteer monitoring brochure):

"The information may be of also be of use to your local environmental organization."

Inexperienced proofreaders usually concentrate on the main text, ignoring the very places where mistakes are most often found-- heads and subheads, photo captions, lists of references, footnotes, and the like.

Typography

Most monitoring publications are produced "in-house," using either word processing or desktop publishing software. Before you invest time and money in desktop publishing technology, ask yourself whether you really need it. A publication done well on a word processor looks much better than one done poorly with desktop publishing.

Whatever technology you are using, watch out for "holes" in justified text, especially when you have a very narrow column. You can end up with this:

The volunteers monitored several parameters, including dissolved oxygen and p   H   .

You can fix the holes if you take the time to insert hyphens, fiddle with word spacing, and so forth—or you can avoid the whole problem by using "ragged right" text (that is, right margin not justified).

If you are a new user of desktop publishing, two helpful books are *The PC Is Not a Typewriter* and *The Mac Is Not a Typewriter,* both by Robin Williams, Peachpit Press, $9.95.

Mind Your P's and Q's

Since government agencies are already skeptical about the scientific credibility of volunteer monitoring programs, it's especially important to use correct scientific terminology. Double-check chemical formulas and spellings of species names, and never write "Ph" for "pH."

Pay special attention to the following three rules. I've seen them violated in quite a few volunteer monitoring publications.

- **It's vs. its.** All you need to remember is that *it's* (with an apostrophe) is always a contraction (standing for either *it is* or *it has*). There are no exceptions to this rule. **Example:** It's been raining for a week. The river has flooded its banks.

- **Quotation marks.** Quotation marks always go outside periods and commas. Again, there are no exceptions.

Example: This is called a "kick-net." *(NOTE:* Different rules apply when quotation marks are used with other punctuation marks, like question marks, colons, or semicolons. Consult a style book for these rules.)

- **Affect and effect.** As verbs, these two words are so widely misused that it's hard not to be confused. To *affect* means to influence or to have an effect on; *to effect* means to bring about.

Example: Our monitoring data obviously affected the board's decision. Our goal is to effect a change in how people dispose of hazardous household wastes.

Writing and Editing Tips

Using the active voice not only makes writing livelier but also helps prevent ambiguity. Passive statements like "data were collected and agency personnel were contacted" leave the reader wondering who did what.

My favorite advice for editors is "never allow something into print that you don't understand yourself" (slightly paraphrased from William Zinsser's *On Writing Well).* Putting this advice into practice is harder than it might seem, but the effort is well worthwhile. A publication can hardly be credible if it isn't even comprehensible.

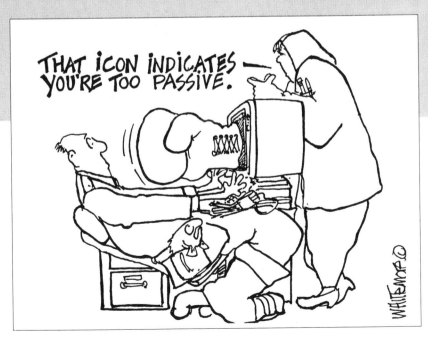

THAT ICON INDICATES — YOU'RE TOO PASSIVE.

PART FOUR
Information into Action

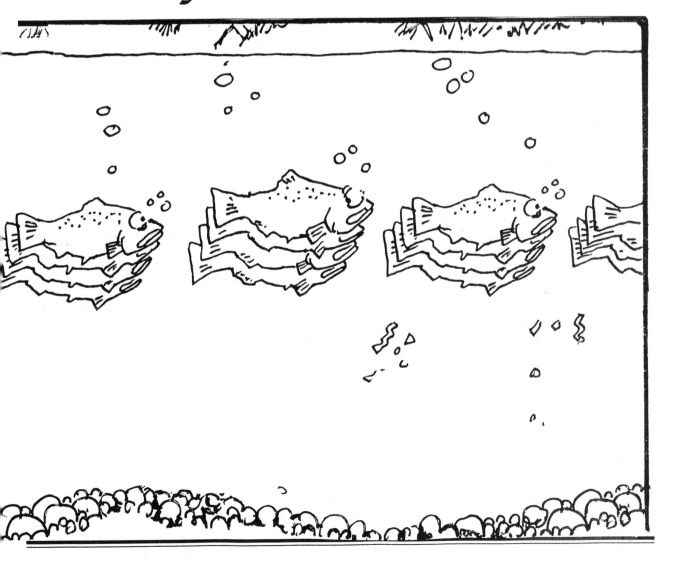

CHAPTER TEN

Streamkeeper Tales

PUTTING YOUR DATA TO USE

The previous chapter posed the question, "Who do you want to impact?" By now you have a pretty good idea of whether or not your stream is healthy. And if your stream has a few problems, you likely have discovered the sources. Now is the time to put your information to use, and hopefully, figure out the cures.

Before you make a decision on which direction to take your data, get to know your target audience. Know what you want to achieve. You also need to know local water quality, stream protection, and land use "rules" (laws, regulations, standards, ordinances, etc.), as well as the "rulers" (elected and appointed officials). As a starting point, take a quick look at the section entitled "Civics 101: A Short and Straightforward Perspective."

Civics 101: A Short and Straightforward Perspective

Individual landowners make the small scale land use decisions that affect your stream. Every property owner likes to be in control of his or her property. However, local, state, and federal **laws or regulations** are established to balance individual property rights with the rights of the public trust.

Federal agencies are tasked with enforcing broad national laws and policies adopted by Congress, and, typically work with state governments and agencies to develop state laws that meet federal standards.

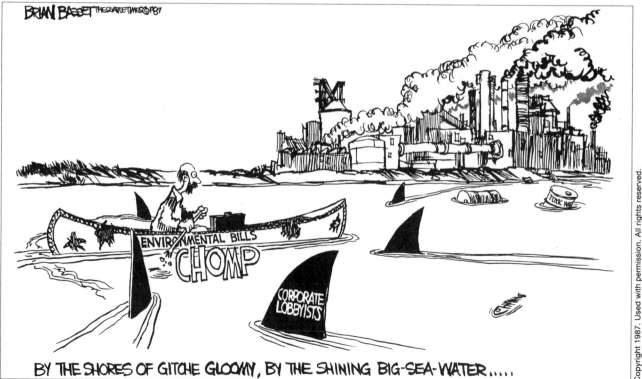

BY THE SHORES OF GITCHE GLOOMY, BY THE SHINING BIG-SEA-WATER.....

The Streamkeeper's Field Guide

State legislatures, comprised of elected representatives, pass state laws which are typically viewed as minimal standards for local governments to follow. Legislators make their decisions after considering a variety of factors. These factors include federal laws and guidelines passed by Congress at the national level, recommendations provided by state agencies, and input from their constituencies and lobbyists. The state legislature also determines the amount of revenues that are provided to state agencies to enforce state laws, as well as the amount of revenues provided to local governments for meeting state standards.

Local governments, which are tasked with meeting or exceeding state (and federal) standards, introduce the rules of law to the public.

Local government planning commissions typically make recommendations on broad land use and water resource policies to **city, town, county, regional district, and parish commissioners or councils.** Planning commission members are usually lay people appointed by councils or other local government legislative bodies. Local government commissioners or council members are also usually lay people, but they are elected by you, the citizen.

Commissioners or councils make decisions on local **laws, regulations, codes, or ordinances** concerning land use and water resources. These decisions are based on planning commission policy recommendations. Commissioners or councils also determine whether or not to provide revenues for enforcing established laws.

Policy recommendations and decisions regarding laws and enforcement are made through a **public hearing** process. Public hearings should be an integral part of your watershed "field" investigations. Good Streamkeeper data presented to local elected and appointed officials during public hearings can be very affective in influencing local laws and policies.

It's up to you to get to know the "rules" and the "rulers." Learn if the rules on the books are being enforced, or not, and if not, why not. Remember that it's the individual property owner who ultimately decides whether or not to follow the rules.

With good, credible information, you can make a big difference on all fronts mentioned in "Civics 101." Remember, the real resource managers are your friends and neighbors and those people you elect to office at the local and state level. The scientists at the agencies are really just advisors who make recommendations. Your information can become more valuable than the scientists if you use it wisely.

The following "vignettes" and tips are based on actual experiences of Streamkeepers. They should provide you with some ideas on how to put your watershed inventory and stream monitoring data into action!

THE BACK YARD - EDUCATING LANDOWNERS

Deer Creek, a very small stream flowing into Puget Sound exclusively through back yards in Edmonds, Washington, captured the attention of Ginny Burger's class at Edmonds-Woodway High School. They sent letters to streamside property owners requesting permission to evaluate the condition of Deer Creek from each yard.

All of the property owners granted creek access to the students, who then analyzed the creek's physical, biological, and chemical conditions. They tabulated their results and developed recommendations on how to enhance the creek. When they sent letters to each of the property owners surrounding the creek thanking them for allowing access to creek sites, the students shared their findings. In addition, the students provided each property owner with a recommendation on how they could improve the quality of the creek's water resources.

The students came up with many recommendations, such as: "Reduce or stop fertilizing your lawn next to the creek, because the excess nutrients cause algae blooms". . . "Replace the grassy lawns which grow up to the stream bank with native vegetation that will provide the stream with shade in the summer, and habitat for wildlife". . ."Make sure that if you construct 'waterfalls' or other features in the stream that you do not cause a barrier to fish migration or destroy existing habitat". . ."Don't dump lawn clippings and leaves in the stream. Yes, they are biodegradable, but they eat up the oxygen supply that fish need to survive."

This project proved to be an effective common sense, almost one-on-one approach that actually reached a vital decision maker, the individual property owner.

THE GARBAGE PROBLEM

During the course of your watershed inventory and stream monitoring exercise, you may have found a lot of trash and garbage in and around the stream. You can use photographs of the debris to convince others that a problem exists. These photos may serve as the "data" necessary to "rally the troops" to get together and do a stream clean-up.

Stream clean-ups are a temporary solution to a more long term problem. Try to work on the

root of the problem - the fact that people are littering or dumping garbage. Before you dispose of the trash and debris you clean-up, rummage a bit to see if you can find some names and addresses of the owners of the garbage you found. You may also want to organize a "Watershed Watch" to keep a look out for the occasional midnight dumper. When it's your turn to be on watch, bring a camera, get license numbers, and call the police when you witness illegal dumping.

Deberra Stecher of Mukilteo, Washington, organized the Picnic Point Creek Protection Association in response to garbage and other problems facing Picnic Point Creek. Deb got twenty friends together with burlap sacks. They collected several tons of garbage and debris. "You know," says Deb, "we got everything from shopping carts to broken toilets. Dirty diapers and newspaper vending machines were popular items. The mattresses and hot water tanks were a real pain, and we made a stack of tires that was ten feet high....we accomplished a lot and felt good, but we don't want to do this again!"

The next week, Deb observed a pickup truck full of garbage backing up near the Creek. Deb took a walk to the truck and advised the driver, "How would you like my shoe in your oatmeal? No? Well our creek doesn't want your trash either. . . Please take it to the land fill and dispose of it properly." The driver, who was a bit embarrassed, took Deb's advice. As he drove away, Deb got his license number for future reference.

STREAM CLEAN-UP TIPS

Before jumping into a creek and removing all debris, litter, logs, and anything else that catches your eye, ask yourself if you are doing more harm than good. All stream clean-up activities should be planned and carried out to help you realize your goal of improving the environment.

Wholesale removal of all materials in the stream bed can actually cause damage. Pools form behind logs and other large organic debris, providing shade and resting places for migrating fish and other creatures. Even human-made material embedded in a stream may be performing a valuable function. When you consider doing a stream clean-up, we recommend that you be very selective in the material you choose to remove. **A simple rule of thumb:** if the object is more than half buried and a great deal of sediment would be stirred up by its removal, leave it there.

Occasionally you may encounter some items that could pose potential health hazards. If you come across 55 gallon barrels containing unknown substances, transformers from telephone/utility poles (these contain PCB's), syringes, or any other suspicious items, DO NOT TOUCH THEM. Contact your local Health District Office or State Environmental Department for assistance.

Before You "Get Your Feet Wet"

Make sure that you have adequate numbers of people to assist you for the area you intend to cover. Begin with a small nucleus of people and hold a few planning meetings. List tasks to be accomplished and the dates they must begin and end. Assign tasks to people at the meetings and follow up on assignments with phone calls.

The following are key tasks to be completed before the stream clean-up:

1. Obtain the property owner's permission to access your adopted stream.

2. Obtain a permit approval from your local Department of Fisheries and Wildlife if required.

3. Arrange for garbage cans and bins at the clean up site. Also ask people to bring trucks and wheel barrows to haul trash and debris.

4. Find temporary locations for storing collected debris near the stream and secure the property owner's permission for that purpose.

5. Arrange in advance for the ultimate destination of the debris collected and the means to get it there. (Your friends' and neighbors' pick-up trucks generally work quite well.) If an entire community or neighborhood is involved, your city or county solid waste departments may provide assistance and free disposal. Contact them well before the event to enlist their help and cooperation.

6. Advertise the clean-up to your community. Send press releases to your local media. You may end up recruiting more help and support.

On the Day of the Clean-up

1. Pick a "staging area" and keep it staffed at all times. Keep first aid supplies, water, and, if possible, provide rest room facilities.

2. Have the volunteers sign in as they arrive. Get names and addresses.

3. Make everyone aware of safety issues, such as: work in pairs, wear rubber gloves, don't pick up hazardous wastes, lift with your legs not your back, don't pick up things that are too heavy, don't walk in the stream in water higher than your knees, etc.

4. Make a map of where people will be working and keep track of everyone involved in the clean up.

5. Make the volunteer effort as festive as possible. Bring refreshments and other treats for when the work is done. Take pictures of the stream and your volunteers before, during, and after the event

Illlustration by Sandra Noel, from *Adopting a Stream: A Northwest Handbook*

6. Keep the day down to about 4 hours.

After You "Dry Your Feet Off"

1. Follow up by recording, in simple terms, the amount and types of debris collected (i.e. truckloads or tons of debris). Take photographs.

2. Provide this information to the media in a press release. This last step will likely result in a well deserved pat on the back for the people who helped in the project and make the public more aware that we should not use our streams as dumping grounds.

3. Send thank you notes to your volunteers.

4. Set up a "Watershed Watch" to prevent more garbage from being dumped into your creek.

THE FARM ANIMAL "EFFLUENT" PROBLEM

During our field investigations we occasionally encounter situations where streams are "running green" with manure from animal waste. During your inventory work you may discover your stream suffers from a enuded riparian zone and excess "effluent" from farm animals. If you happen across a stream reach that flows through a feed lot full of 200 head of cattle with free access to the stream, you will have a fair idea of the magnitude of the problem.

In this situation, the following scenario is probable:

You have monitored a stream reach upstream and downstream of the "alleged" problem area. Your data clearly demonstrates that there is low dissolved oxygen, high temperature, and high levels of fecal coliform, phosphorous, and nitrogen in the downstream site and the opposite conditions in the up-stream site.

Your analysis of the macroinvertebrates in the upstream site found a diverse and abundant population of intolerant species. Only a few worms were found at the downstream site. Trout were found upstream, no fish were observed downstream. You have photographs of "contented cows" depositing "effluent" into the stream. Your photos also document that the water is clear in the upstream site, and "running green" at the downstream site.

So, what do you do? Here's a multiple choice list for you to choose from:

a. Call your local state agency responsible for enforcing water quality standards, describe the problem and provide your data for their use.

b. Share your information with the media.

c. Contact your local Natural Resources Conservation Service office and describe the problem. Provide your data for verification, and request that they work with the land owner to develop a "Best Management Practice Farm Plan" that will fence out the farm animals, provide a localized watering spot for the animals, and revegetate the stream banks.

d. Find some fellow Streamkeepers who are willing to work a few weekends installing fences and planting trees and shrubs.

e. Locate sources of funding to pay for fencing materials and plants. Two examples of funding sources include: the Natural Resources Conservation Service's cost sharing program and the U.S. Fish and Wildlife Service's stream and wetland protection funds.

f. Contact the landowners. Listen! Learn their attitudes about streams and fencing, and whether or not he or she is willing to change the situation. Share your information with the landowner in a non-confrontational fashion.

We have found that starting with "f" is always the best approach. If you listen, you will find out that most land owners are not intentionally creating a situation designed to pollute. For example, the landowner above may say, "I let the cattle have the run of the

I HAD A THOUGHT LITTLE DOGGIES! LET'S BYPASS THE STREAM — MINIMIZING THE DAMAGE TO THE RIPARIAN ZONE.

AN EARLY ATTEMPT AT GREEN CORPORATE POLICY

creek because they need the water." This will give you an entry point to suggest fencing the cattle away from the creek except from localized watering points which are "hardened" with gravel to prevent bank erosion. Or, better yet, if gradient allows, suggest a gravity fed watering trough that is away from the stream entirely.

If you can throw in the possibility of "d" and "e," money, materials, and a labor force to install fences and plant streamside vegetation, you might catch the landowner's ear. Add in the possibility of having a farm plan developed as suggested in "c," and you may move forward or backwards depending on what you learn while "listening."

Calling in the "officials" as suggested in "a" should be kept in your hip pocket as an avenue of last resort. Perhaps the landowner will like the idea of "getting those bureaucrats off my back" and let you move ahead with "d" and "e" so they can avoid the government agencies.

On the other hand, if "c" through "e" don't work, pursue avenue "a" with vigor. And if that does not work, don't be afraid to bring in "b," the media. By then, you will have a dossier of information that should be of interest to media and can help rally support needed to correct the problem.

THE FISH MIGRATION BARRIER

Suppose you have conducted your watershed inventory and found that your adopted stream flows through a culvert under a road. On the downstream side of the culvert there is a four foot drop to the stream bottom.

You have checked state fish and wildlife department records and found that silver (coho) salmon spawn and rear in this stream, but that they have been blocked from migrating upstream ever since the culvert was put in. You checked the date stamped above the culvert and discovered that this problem started in 1939. You checked your watershed maps and, using your string measuring device compared to the map scale, determined there are nine miles of spawning and rearing habitat upstream.

You set up a monitoring station at the culvert and acquire great data on the volume and velocity of water flowing through the culvert. You compare this information with the "burst speed" capabilities of silver salmon (found on Table 6, page 114.) Your data confirms what your eyes and historical information tell you. This culvert is a barrier to salmon migration.

Your data also provides you with the information needed to design a fish ladder to correct the problem. You know the proper distance between "steps," and the desired pool depth within each step, by comparing the ideal situation as presented in the *Stream Enhancement Guide*, published by the Ministry of the Environment, British Columbia, Canada, and the flow and gradient information you have collected.

During your information gathering process, you find that the property adjacent to the road and culvert is in a public right of way, and that the culvert was installed by county road crews, but in 1939 no one was too concerned about salmon migration. Remember, this information is data. Data includes facts or information to be used as a basis for discussing or deciding something.

You contact the local fish and wildlife habitat manager and find out that the state is requesting counties to remove all barriers to fish migration caused by county roads. You dig a little deeper and found that your county doesn't have the budget to prepare a design and pay crews to build a ladder.

You contact your county's public works director, who is in charge of the road crews, and find out what kind of resources are available. In advance of the meeting you are able to locate a labor force willing to do the work, such as a local conservation corps group. (You will find that there are all kinds of volunteers available to do the "grunt work," just not too many able or willing to take the lead to plan and coordinate projects).

You put all of your data on the table, and inform the public works director that she or he is responsible for removing the fish migration barrier, but that you have a design and labor force ready to go to work. You just need a little help. Because of all your efforts, the county is able to provide you with that help.

This scenario actually happened on a small unnamed creek flowing into the Snohomish River in Washington State. Teachers and students from Sunnyside Elementary School and Marysville High School got the Snohomish County Public Works Department involved. The county supplied trucks to deliver donated construction materials to the site. A volunteer labor force was put into action, and in just four weeks, a fish ladder was constructed and operational with no out of pocket expenses. That fall, those involved in the project returned to the creek to witness salmon above the new fish ladder. Another success story.

FISH LADDER

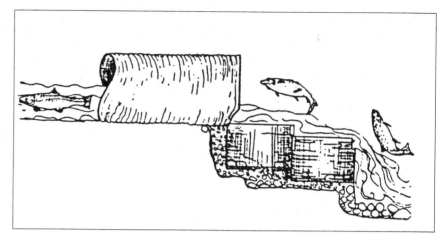

THE DIRT PROBLEM

Perhaps your watershed inventory determined your creek was in pretty good condition. The water chemistry fell within acceptable limits, the temperature was reasonably constant within the seasons, the flows were regular and the fish population was productive.

On a nice rainy day, you were on one of your quarterly site visits to sample the macroinvertebrate population. Much to your surprise you found that the water, which had always been crystal clear, was running brown with sediment.

This situation led you upstream where you located a five acre area next to the stream that was recently cleared of all vegetation. There were no erosion or sediment controls on the site, and sediment was pouring into the creek. There was no one on the property.

You moved back downstream to your monitoring site and checked the macroinvertebrate situation. You found that the intolerant species you were expecting were dead.

What should you do next? In this kind of situation, immediate action is necessary. If this scenario took place during fish spawning season, an entire year class of fish could be destroyed in a few hours, their eggs smothered with sediment.

The best course of action in this situation is to contact your state fish and wildlife department (in Washington State, you can contact a fisheries or wildlife patrol agent by contacting the state highway patrol). Describe the situation to the officials and provide exact directions. By sharing your observations (data) on sediment and the demise of the macroinvertebrate population, you are likely to get a fast response.

It is possible that a variety of local "rules" have been violated by the property owners of the cleared site. From your watershed inventory, you will have a pretty good idea what the "rules" are, and know who to contact for enforcement action.

If the property owners or contractors are on the site, you could inform them of the problem. (If they are not present, you may want to try to find out who they are and arrange a meeting with them). Even if you suspect that they are aware of the problem, give them the benefit of the doubt - find out if they are simply unaware of the rules rather than knowing violators. Suggest to them, in a non-confrontational way, that they take immediate steps to correct the problem. If they take no action, then call your local fish or wildlife officials as mentioned above. Also remember to enlist the help of your local media.

THE THREATENED URBAN HEADWATERS

Picture that you have completed an inventory of your watershed. You have located all of the aquatic features that the stream relies on for summer flows. You are familiar with the comprehensive land use plan prepared by the county planning department for the area that the stream flows through. The plan's policy statements calling for protection of streams and wetlands pleases you.

You have also looked over the zoning maps for the land within your watershed and found that the headwaters are currently zoned for residential use. A closer look at the comprehensive plan reveals that the headwaters are designated for commercial use in the future. You detect the inconsistency between the stated policies about protecting streamside wetlands and the commercial land use designation.

During your continuing monitoring activities of the stream, you see a notice tacked up on a telephone pole. The notice announces a rezone proposed for the property surrounding the headwaters of the stream. The rezone would change the zoning of the property from single family residential to commercial.

When you inquire about the notice at your local planning department, you find out that the landowner has requested a hearing in front of the hearing examiner to request the zoning change. The landowner plans to put the entire site (5 acres) into use for commercial purposes.

From your inventory work, you discovered that wetlands occupy three of the five acres on the site. You quiz the planning department about the comprehensive plan policy that protects streams and wetlands and their map showing a commercial area imposed on top of a wetland. The planning department acknowledges that they failed to recognize the wetland when they put the maps together. They tell you that the hearing examiner will have to make a decision on the case.

You have a lot of data. You have accurately mapped the stream and its headwater wetlands. You have historical and new information on salmon and trout runs dependent on the stream. This should be a "slam dunk" for your Streamkeeper group, right?

Wrong. The property owner makes a strong case about investing capital based on a comprehensive plan designation approved by the county council. The landowner has hired engineers who state that the site has some wet soils but is not a wetland, and that there should be no problems with construction.

County staff recommend that the hearing examiner deny the rezone request because they feel the area is a wetland. They acknowledge the error in mapping; staff biologists testify that the area is a wetland and should not be filled.

The hearing examiner now has conflicting testimony.

The floor opens for testimony from the public and the local Streamkeepers are ready. Not only have they documented the existence of the wetland with maps, they have made a video tape of the site as well. One scene that carries the day shows a Streamkeeper on the property immediately downstream from the site (nicely avoiding a trespass allegation) with a stadia rod that shows the water is three feet deep.

The Streamkeeper narrates the video by saying, "this site is a palustrine emergent wetland that serves as the headwaters of Cemetery Creek". . . the camera then "pans" to the background . . . "water lily and duckweed are found in the open water area, and as you can see, the rest of the wetland area is dominated by cattails, skunk cabbage, and willows. All these plants are characteristic of a wetland."

Compelled by the video which supports the biologist's claim, the hearing examiner denies the rezone.

WHITEHEAD© WITH APOLOGIES TO BRIAN BASSET

LAND USE IN THE RIPARIAN CORRIDOR

You met Debbera Stecher and the Picnic Point Creek Protection Association in "the Garbage Problem" tale. In addition to organizing several stream clean-ups, and a "Trash Watch" that has caught several midnight dumpers, the members of the PPCPA have also collected information about the resident trout population in their small urban stream, and kept records on the wildlife within the watershed.

Futhermore, the PPCPA investigated the property ownership, the comprehensive land use plan, and all of the land use zoning within the watershed. They located an aerial photograph map of their watershed that was of a scale where one inch equals 400 feet. The photograph was taken in 1985. All development that had occurred in the watershed up to that date was highlighted.

Several new developments had occurred in the watershed since 1985. And several more were contemplated in 1992 and 1993. The PPCPA was concerned that the cumulative impacts of these proposed new developments were not being evaluated properly by the elected officials responsible for decisions on stream buffers and wetland protection.

The PPCPA presented their concerns at a "Conditions of Approval" hearing presided over by the county council. At issue was a new development proposal which would remove significant areas of riparian habitat. Debbera presented the PPCPA's data on the physical and biological characteristics of the stream. She pointed out the concerns that her organization had with the proposed design of the new development. She also utilized her aerial photograph map very effectively.

Specifically, the PPCPA presented the council with a picture of the Picnic Point Creek Watershed as it appeared in 1985. Then they added a plastic overlay demonstrating all of the recent and proposed new development in the watershed. They asked the council to consider evaluating the cumulative impacts of all of the new development, rather than looking at each "with blinders on."

The PPCPA requested that no new development occur within the watershed until a basin-wide drainage plan was completed, and that the riparian habitat encroachment proposed by the development be put on hold until the drainage plan that considered all new developments was complete.

The council decided against requiring a basin-wide drainage study as a requirement for the development's approval. However, because of the data presented by the PPCPA, the council did establish a number of "conditions of approval," including preservation of the riparian corridor, redesign of all lots adjacent to the creek, designation of a "native growth protection area," and removal of proposed roads that would have infringed on the riparian corridor.

THE LOST SALMON RUN THAT CAME BACK

In 1983, the teachers and students from the Jackson Elementary School adopted Pigeon Creek in Everett, Washington. Fifth grade teacher, Brandon King, remembered from his youth that "salmon were so thick in the creek that you could walk across their backs." Unfortunately, strip development sprang up in the headwaters of the creek in 1958, resulting in increased volume and peak flow of water during the winter months, eroding stream banks and smothering spawning gravel with silt.

The salmon runs disappeared. The Jackson School established a long term goal to restore the salmon run to Pigeon Creek. They put together a long term (20 year) action plan that included conducting a watershed inventory and monitoring the health of the stream. Students also became

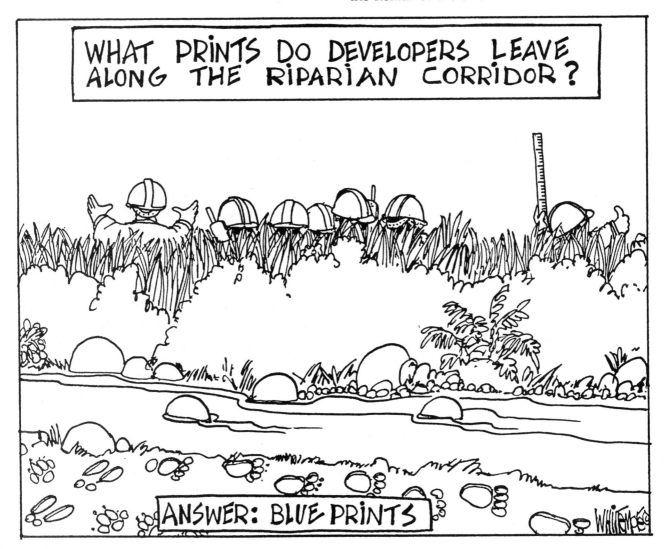

the ambassadors for the creek to the community. They produced a watershed map showing the creek and several land use features. They reprinted a flyer from the Adopt-A-Stream Foundation and the Washington State Department of Ecology entitled *Streams: Guidelines for Survival*. The students distributed their map and the flyer to every household and business in the watershed with a plea to read the information and help restore the salmon run.

The students initiated a variety of other activities. They conducted stream clean-ups, marked storm drains that flowed into the creek with the message "dump no waste, drains to stream," and released salmon they raised in their classroom aquariums into the creek.

The students also put together a slide show of their activities and made a presentation to the Everett City Council. They closed their presentation with a request that the city consider addressing the stormwater runoff problem that led to the demise of the salmon run.

A group of teachers, parents, and concerned citizens decided to help out, and formed the Everett Streamkeepers. They began to monitor all new land use proposals in the watershed. Their diligence picked up on a major subdivision proposal that would encroach upon the riparian habitat of Pigeon Creek, and route all stormwater runoff directly into the creek without filtration.

The Everett Streamkeepers presented to the city council a more environmentally sound, alternative design for the proposed development. As a result, the final project featured vegetation buffers and stormwater biofiltration strategies. The subdivision is now called "The Preserve," and its more environmentally conscious design ended up being very profitable for the developer.

In 1987, the first salmon in more than 28 years returned to Pigeon Creek. The city began designs for a stormwater control facility to be constructed in the head-waters. And the Jackson School activities continued, involving the entire student body and faculty. . . and then almost got buried alive.

FINALLY THELMA THOUGHT SHE HAD FOUND HER BLISS...

THE STREAM THAT ALMOST GOT BURIED ALIVE

Jackson students, the Everett Streamkeepers, and Pigeon Creek had still more adventures to face. The Port of Everett submitted a proposal to the U.S. Army Corps of Engineers to encase the lower end of Pigeon Creek into a culvert 400 feet long where it flows into Puget Sound, cover it with fill and construct a facility over the top. In the wetlands filling permit submitted to the Corps (required under Section 404 of the Clean Water Act), the Port did not mention Pigeon Creek at all by name, referring instead to a "drainage pipe" that flowed into Puget Sound. Pigeon Creek is culverted for a short distance right at its mouth to flow beneath railroad tracks; However, most of it (except for road crossings) is above ground.

By keeping up their monitoring activities, the Jackson School students and the Everett Streamkeepers were able to provide data during the permit review process to demonstrate that Pigeon Creek historically was a viable salmon stream, not just a "drainage" system. They were also able to demonstrate that the creek was in a state of restoration, and that salmon were returning once again.

The 404 permit was denied. The creek still flows above ground from top to bottom! In 1992, the City of Everett celebrated the completion of a stormwater control facility that has already begun to cure the problem with the salmon spawning beds.

Thanks to the way the Jackson Elementary students and the Everett Streamkeepers used their data in this case, the future of Pigeon Creek and its salmon is bright.

THE BURIED STREAM THAT WILL SEE DAYLIGHT

Under the leadership of Kit O'Neill, the Ravenna Creek Alliance in Seattle, Washington compiled a wealth of data that includes historical, physical, biological, and fiscal infor-mation on their local Ravenna Creek.

Kit put all of this data into a great document entitled *Ravenna Creek: Past, Present, and ?Future?*. The opening chapter, "The Historical Dimension," compares an 1894 Seattle road map with an updated 1908 USGS map, showing dramatic changes. The complementing text brings the reader through the decades, up to 1948 when the city diverted the creek into a trunk sewer line.

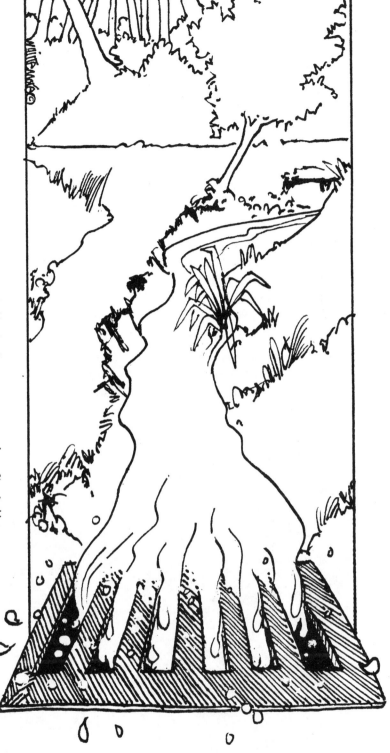

The study then brings the reader into the present, pointing out positive features of the portion of the creek which is still above ground, in the City of Seattle's Ravenna Park. It describes abundant wildlife in this small oasis of 69 acres, and a small resident fish population in the one mile long channel of open water. Unfortunately, Ravenna Creek literally "goes down the drain" at the downstream end of the park, where it is connected into the municipal sewer system.

With minor filtration, the approximately 800,000 gallons of water a day that flows into the sewer system would be suitable for drinking water. Unfortunately, the city's tax dollars are paying to have this water treated as sewage. It also contributes to sewer overflows into Puget

Sound during periods of high rain (which happens often in Seattle).

The Ravenna Creek Alliance has successfully presented the fiscal impacts of this situation to "Seattle Metro," the entity providing sewerage to the Seattle metropolitan area. Metro has decided that the water from Ravenna Creek should be diverted away from the sewer system and back to where it used to flow (Lake Washington). Metro has calculated that continued treatment of the creek flows will cost more than $2,000,000 over the next twenty years. It has also figured that it will cost approximately $800,000 to connect the creek back into the lake via a pipe system.

The Ravenna Creek Alliance has presented a variety of surface flow design concepts for the creek as an alternative to the pipe approach, and has convinced Metro to devote the funds for the pipe approach to a restructured streambed where Ravenna Creek will flow above ground, in the daylight, through a residential and commercial area on its way back into Lake Washington. Thanks to the thoroughness of their preliminary feasibility study and development of several design alternatives, Ravenna Creek has a chance to "see daylight." The Ravenna Creek Alliance is using its data to form financial partnerships with a variety of public and private interest groups who collectively will be involved in property acquisition, construction of habitat features, and long term maintenance of the "resurrected" Ravenna Creek into the future.

THE OIL AND WATER PROBLEM

John Cronin is the Hudson River Keeper. He is not connected to a government body or industry. With the help of a lawyer and nine law student interns, John gathers data to enforce the Clean Water Act (CWA) passed in 1972. This federal law called for "clean, fishable, drinkable water" nationwide by 1977 . . . a goal that, for some water bodies, we have missed by a few years.

Besides setting national standards and levels of treatment for water pollution, the CWA gives

FROM SEA TO SHINING SEA TO SHINING SEA

private citizens the right to act as their own private attorney and take polluters straight to federal court. John uses this section of the law as a tool to put his data into action. Any Streamkeeper can do the same.

The first polluter John tackled was an oil tanker that moved up the Hudson River, flushed out chemical residues from its tanks, and then moved further upstream to fill its tanks with fresh water. It then traveled south, selling the water to Caribbean nations. John staked-out the tanker and caught the ship's crew in the act.

Because John had good data from previous stake-outs, he was able to convince a CBS-TV camera crew to join him on what turned out to be the last time the tanker flushed its tanks into the Hudson. CBS captured the polluters on national television. The problem was stopped, and a huge, well deserved fine was slapped on the violator.

Building on that success, the Hudson River Keeper has been collecting data on a variety of other polluters throughout the Hudson River Watershed. This data has led to individuals, businesses, and cities and towns correcting problems that they are causing, and a tremendous improvement in the quality of the Hudson River's water resources.

Note: Congress has weakened the provision of the CWA which allows citizens the right to sue polluters.

THE SIXTH GRADE APPROACH TO SHARING DATA WITH A TOWN COUNCIL

Lets assume you are a bit below voting age, but you have completed a watershed inventory, and discovered quite a few problems facing your stream. Then you discover a few human activities planned in your watershed that may make matters even worse. What do you do? Present your concerns and supporting data to the decision makers. Following are two great examples of how some young Streamkeepers handled similar problems with slightly different approaches.

West Coast

Jeff Swanson's sixth grade students from the Post Middle School in Arlington, Washington,
adopted Portage Creek. They discovered that this small tributary of the Stillaguamish River had its own distinct watershed, encompassing 18 square miles. They discovered that it was the home to king, pink, and chum salmon as well as steelhead trout. They also discovered that nonpoint pollution had taken a taken a serious toll on the stream's fish and wildlife populations.

The sixth graders were keeping an eye out for any new problems that might affect the stream. They noticed some signs on telephone poles indicating a "proposed land use" action. After a couple of phone calls to Arlington City Hall, they discovered a plan to build 20 new townhouses right next to the creek. So, the students paid a visit to City Hall. They requested to see the construction plans for the proposed housing project and discovered a few design flaws that could cause additional problems for the Portage Creek's already depressed salmon run.

Armed with watershed inventory data they had been collecting over the previous six months, the students became very effective participants in the City Council's public hearing on whether or not to approve the development. The chief spokesman, sixth grader David Ray, was supported by classmates Bryan Gregory, Tuffy Miller, and Ian Somes.

David presented their inventory findings, some existing problems in the creek that they discovered, and an outline of their plans to correct some those problems. He said, "We put together a plan to clean out garbage and litter, plant deciduous trees for shade and shelter, remove silt from spawning areas, build habitat structures to increase pools and riffles, and to create signs near the creek and at storm drains to make people aware of Portage."

David's testimony went on to say that, "Protecting the creek is not a short term project. We are committed to watch over it until our grandchildren are ready to take over. This is a huge job and we will be looking for help from several agencies and the entire community."

Ian Sommes added, "There's only one Portage Creek and we have to save it." Then David followed up with specific recommendations that the students hoped would become conditions for approval of the proposed new development. Speaking for the students, David requested:

> "There should be a buffer zone of 50 feet on either side of the creek to reduce soil erosion and runoff pollutants. When an area is being developed, erosion control fences should be built to reduce sedimentation, grading of land should only occur in spring or summer, and grass should be planted immediately after the grading is done. The number of culverts should be limited and they should be 'bottomless half culverts' so they won't disturb the sensitive spawning areas. Water running into the creek from the development should be filtered through wet ponds and biofiltration ditches. These are needed to reduce flooding and will filter out residential pollutant such as oil, fertilizers, detergents, and antifreeze."

David concluded his testimony by saying:

> "It is possible with the help from you and others who make regulations, to save Portage Creek and the salmon who live there. Water quality standards must be made with the long term in mind. There is a lot of land out there that can be developed, but there is only one Portage Creek. We want the salmon and trout to return to a healthy habitat. Without your help, it may be impossible to save Portage Creek which is already in poor condition and is not improving. With your support we can turn things around."

Needless to say, the Council was impressed. They directed the developer to incorporate the students' recommendations and to present a new site design *to the students* for their approval!!

The developer, Steve Warren, met with the students and appealed to them to avoid the bottomless arched culvert, citing its $30,000 cost as being excessive. The Students disagreed, saying, "Protection of the creek is worth it." After meeting with the students, Steve again met with Council and said "I plan to make this (the planned development) a showpiece. I will work with the city staff, the Department of Fisheries, and the sixth-graders in Jeff Swanson's class and do whatever it takes."

The amended project got the Council's approval, and both the developer and the students were able to celebrate.

(This tale contains part of an article from the *Arlington Times*, Arlington Washington.)

East Coast

In Yorktown Heights, New York, John Holland's sixth grade students at Mildred E. Strang Middle School have adopted Hunterbrook Stream. His students faced a situation very similar to the their West Coast counterparts. A proposed development was going to encroach on the riparian corridor of Hunterbrook Stream, and little consideration was being given to the environmental impacts. John's students tackled the situation in a different fashion.

The Yorktown Heights Town Board conducted a public hearing to hear comments on the development proposed next to Hunterbrook Creek. The hearing was broadcast on Public Access television. Using procedures from a draft version of the *Streamkeeper's Field Guide*, John taught his students how to collect and analyze macroinvertebrates, and how to understand the relationships between macroinvertebrates, clean water, and fish. The young Streamkeepers decided to use their five minutes of testimony to teach the Town Board these relationships.

John said, "The students gave macroinvertebrate sorting sheets and a variety of alcohol-preserved macroinvertebrates from one of our test sites to town officials. They showed the Board members how to group the samples according to pollutiontolerance. The sixth graders waited patiently and offered assistance to a couple of officials who were having trouble differentiating mayfly larvae from stonefly larvae. The interchange following the identification 'quiz' was truly education.

"The Board learned that removal of the riparian vegetation, and with runoff from the proposed development would adversely affect the macroinvertebrate population, and subsequently the good fishing in the stream would disappear as well." John went on to say, "the developers who were on the evening's agenda declined to speak that evening. The Board subsequently denied the proposal 'without prejudice' (meaning the developers could redesign and resubmit), and directed the developers to prepare a more environmentally sound proposal. My kids were elated!! They had real political clout."

THE STREAM WITH NO NAME

The locals called it the "Gully." The small stream, which flowed through the middle of Blaine, Washington, near the Canadian border, was treated that way. Western Washington University student Brad Vinnish decided to make some changes.

After graduating from Streamkeeper Field Training, conducted by the Adopt-A-Stream Foundation, Brad trained teachers from the local middle and high schools how to inventory and monitor the Gully and its surrounding watershed. Jim Jorgenson, Blaine High School teacher, said, "Our students found that many areas of the stream had been 'trashed' and even put underground in pipes. The kids got together and organized clean-up efforts on many of those areas that were really a mess." The students collected a variety of physical, biological, and chemical information on the creek.

The local media picked up on these activities and Brad, the teachers, and other students found that there was some positive interest from the community in taking care of the creek. The kids got together and requested an audience with the City Council. Their goal was to make the local elected officials aware that the Gully actually had a small population of trout, and with proper care, could once again become a salmon stream. They also wanted the city to recognize officially that the creek existed, and that it was not just a "drainage ditch" or "gully."

Brad and his fellow Streamkeepers also had done their homework on another topic. They found that there was an official State Board of Geographic Names administered by the Washington State Department of Natural Resources. They contacted this agency, asked and received an application that could be used to give the Gully an official name.

At the next city council meeting, which was attended not only by Streamkeepers but the media as well, the Council was shown all of the data that the students collected. The Council was also asked to consider recommending a name for the creek that could be submitted to the State Board of Geographic Names.

The Council responded positively. They decided that the creek should be named after the original homesteaders of the area, suggesting that the stream become "Caine Creek." The recommendation was approved at the State level, and the Gully officially became Caine Creek. Now the stream appears as Caine Creek on maps. Stream identification signs have sprung up at road crossings in Blaine.

It is interesting what can happen when a stream gets a new identity. The stream naming process in Blaine has since led to an ongoing community wide stream restoration effort. The Council is even considering "daylighting" parts of the creek that have been put in culverts.

THE FUNERAL DIRGE - OR- WHEN ALL ELSE FAILS

Mary Anne Tagney Jones lives in an unincorporated part of King County, Washington near Seattle. The King County Council was amending its land use plan and held a few public hearings to see what the public wanted. In advance, the Council posted notices on some phone poles, but the word didn't really get out until it was decision time. The Council's last hearing was to render a decision on land use in and around Mary Anne's unincorporated community. Those decisions could affect the long term health of the area's streams.

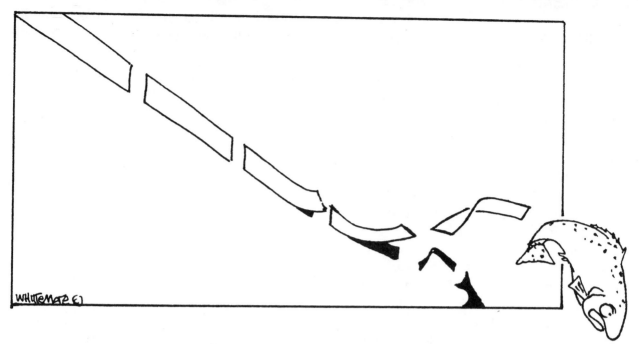

Mary Anne appealed to the Council to allow one more date for receiving testimony. They refused. First, Mary got angry; but then, she got creative. When she organized a small community gathering, she found that others were also unaware of the pending land use plan change. The community decided that it was going to be heard.

In order to get their message across, Mary Anne's group became rather unconventional. On the evening of the hearing, they waited until the hearing room filled up with people. Then Mary Anne and her band of creative fellows, all dressed in black, walked down the aisle of the hearing room toward the "rulers."

They were carrying a cardboard casket with the name of their community on the sides, and handing out to the audience a well prepared statement expressing their concerns on the land use issue at hand...all the while humming a funeral dirge! After providing the last hand outs to a mute council, Mary Anne and her now merry band, left the hearing room. Outside, Mary Anne's group had an opportunity to present its views to three television networks who had received prior word on the "mock funeral."

As a result of this Streamkeeper group's appearance at the public hearing, the Council had a change of heart. They accepted Mary Anne's "data" and incorporated them into the land use plan.

We recommend you undertake this type of approach only after you attempt more moderate or conventional approaches. However, if your stream is in jeopardy, and conventional approaches do not seem to work, "playing hardball" with something that is unconventional may be more effective.

As you can see, there are many ways that Streamkeeper information can be put to use. As you become more familiar with your watershed, you will discover that you have many allies and that help is often just a phone call away.

InWashington State, for example, if you find a major problem in a stream (like a bulldozer without a permit) you can alert the State Department of Fish and Wildlife Patrol agents by calling the State Highway Patrol radio dispatch. Most local, state (provincial), and federal resource management agencies have "hotlines" that deal with issues ranging from grading permit violations, to poaching, to water pollution.

Also learn the names and phone numbers of your elected representatives at all levels. They should be working for you and your watershed. If they aren't, use your data to show them the way!

Closing Thoughts

Tom Jay is a poet and artist who heads up an organization called Wild Olympic Salmon located on the Olympic Peninsula of Washington State. He has a wonderful way of getting at the "root" of words in his poetry. Once we had the good fortune to hear Tom describe the word ecology. His description bears repeating:

"Picture a pair of adult salmon returning to your stream in the fall. The female will dig out a nest or redd in the gravel and deposit eggs which are then fertilized by the male. Within a few days the adult salmon will die.

Their bodies become food sources for a variety of wildlife, and if they decay in the stream, nutrients from their bodies become a source of energy for the surrounding watershed. In fact, if you were to do a radio isotope study of alder trees along the banks of the stream, within two weeks you would discover some of the nutrients from the decaying salmon have been absorbed into the trees' root systems.

Over the next hundred days or so, depending on the water temperature, oxygen levels, and flow characteristics of the stream, the salmon eggs will hatch, and the baby salmon or alevin will use up the food from their yolk sacs. Emerging as salmon fry, these juvenile fish now have to seek out food on their own. This time frame coincides with the 'spring hatch,' a time familiar most anyone who has put a fly rod to use.

The spring hatch is when underwater insects like mayfly, caddisfly, and stonefly emerge from the water as adults. They sprout wings and fly up looking for mates. After mating, some of these insects will light on the alder tree which is now sprouting leaves that were, in part, nurtured by the nutrients from the adult salmon which died

last fall. These insects then die and fall into the water, becoming food sources for the baby salmon.

As autumn returns to the stream, the leaves from the alder trees will fall from the branches and float down to the stream. The leaves will sink to the bottom of the stream where they become sources of food for juvenile insects bred from the winged adults that mated on the leaves during spring.

The baby salmon feed on insects that fall from trees. The trees are nurtured by decaying salmon. And the insects are nurtured by leaves from the tree.

As you can see, everything is interrelated . . . connected together. The true meaning of the word ecology."

The Adopt-A-Stream Foundation often shares this story with potential Streamkeepers to emphasize the need to understand the biological integrity of the entire watershed that a stream flows through. At the same time, we point out that streams are part of the public trust.

As explained by Phillip Wallin, Oregon River Network Director, the public trust doctrine, adopted by the courts of most states, says that "the public has the right to use the flow of rivers and streams for fishing, boating and other purposes. Most importantly, it says that this right predates and supersedes the rights of private users. Streams and rivers, like the ocean beaches, do not belong to private parties. Governments may grant rights to private parties for various uses of water and streambeds, but these rights are limited by the prior right of the public."

No one has the right to pollute a stream or destroy its natural systems. No one has the right to destroy an ecosystem under the guise of exercising personal property rights. Yet when we look around, our streams and rivers are being terribly abused. The public trust is under assault.

"GRAMPS, ARE YOU SURE THIS IS WHERE YOU USED TO SNAG THOSE SPECKLED BEAUTIES?"

A 1982 Nationwide Rivers Inventory found that fewer than 2% of streams in the contiguous 48 U.S. states remained at "high natural quality." More recent studies from a variety of sources show that more than 50% of America's wetlands have been altered or destroyed.

In a 1993 presentation to Congress, Dr. James Karr of the University of Washington Institute for Environmental Studies reported the findings above and these additional disturbing bits of information, researched by Dr. Art Behnke:

"Only forty-two American rivers longer than 200 kilometers have not been dammed. Sixty to eighty percent of our riparian corridors have been destroyed. Mink that are fed salmon from the Great Lakes don't reproduce. Fresh water macroinvertebrate populations that fish populations are dependent on are becoming rare and endangered in the majority of our river systems."

We should all be outraged by this invasion of the public trust and the alarming results. Sports fishing is on the decline, the commercial fishing industry is in trouble nationwide, many outdoor recreation opportunities are disappearing forever, and safe drinking water is becoming more difficult to find.

Looking at the results from an individual fish species' point of view can be even more alarming. For example, 100 of the 400 pacific salmon stocks are now extinct and 214 others are considered by the American Fisheries Society to be at moderate or high risk of extinction. And other species of fish are suffering a similar fate in watersheds around the country. In fact, at least 40 species and subspecies of U.S. freshwater fish have become extinct because of habitat alteration, introduction of exotic species, pollution, or overfishing.

By reading this book, Streamkeepers will be armed with tools to gather information that can be used to make positive changes in their watersheds . . . information that should help to turn the tide on some of the unsettling facts described above.

As you embark on your Streamkeeping efforts, remember what we have emphasized throughout the course of this book: that you cannot evaluate the quality of your water resources simply by taking chemical measure-ments. While water quality from a chemistry perspective does provide you with useful information, a Streamkeeper should evaluate the **biological integrity** of an entire watershed.

Thus, in addition to monitoring **chemical water quality**, Streamkeepers must also become knowledgeable of the **habitat structures** of their streams in order to understand what kinds of fish and wildlife their watersheds support. Woody debris, rocks, pools, riffles, and varied stream bottom compositions are all biologically intertwined.

Streamkeepers should also know about their streams' **energy sources**. Streams get energy from decaying salmon carcasses as suggested by Tom Jay, as well as from leaves, fallen trees and other natural organic matter. They also receive energy of another kind, from fertilizers and sewage produced by humans. Streamkeepers are challenged to find the balance between the natural and human-induced energy sources that their watersheds can stand.

When fish runs disappear, the typical human reaction is to treat symptoms instead of curing the problem. Rather than protecting habitat that indigenous species of fish require, we humans often try to introduce exotic species that can adapt to the changing environment. We have a tendency to build bigger and bigger hatcheries with the misguided notion that pumping millions of fish into a stream will make up for the loss of habitat. We have failed to recognize the **biotic interactions** in our watersheds. We have failed to learn what Tom Jay pointed out, that everything is interconnected.

Water chemistry, habitat structures, energy sources, and biotic interactions are all linked together and affected by the **flow regime** of your stream. Perhaps this is the easiest of all the factors for Streamkeepers to understand because changes in stream flows are easily detected when people replace open areas, wetlands, and riparian corridors with shopping centers, rooftops, and parking lots.

By understanding the relationships between all five of these factors, Streamkeepers will have a holistic view of the biological integrity of their streams. To put that new found knowledge to work and to affect positive changes to our water resources, Streamkeepers must also have a strong understanding of a sixth element, the **"Human Factor."**

Find out who makes the land use decisions that can affect your stream's health. Understand that the broad federal, state, and local laws designed to protect your stream may not be enforced unless you are continually monitoring the **political integrity** of your stream

at the same time you are monitoring its biological integrity. Realize that you have the power to ensure that these laws are enforced, and to change them if they are not working.

Accept the challenge to become a Streamkeeper. Take action to protect and enhance that stream "in your own backyard." The action you take may be something as simple as doing a stream clean up or putting up a stream identification sign, or as complex as putting together a dossier of physical, biological, and chemical information about your stream and initiating a law suit to protect it.

Don't sit back and let others do it for you. . . it might not happen. Take responsibility for the future and defend the public trust by becoming a Streamkeeper. It could be the most rewarding experience of your life.

Appendices

Reprinted, with minor modifications, with permission from the Oregon Department of Fish and Wildlife, from *The Stream Scene: Watersheds, Wildlife, and People.*

Hand Seining Sampling Methods

Data collected with the hand seining method must be done using consistency with technique in the same areas over time in order to produce reliable information.

1. Choose a representative sampling section within the survey area. Different species of fish use different parts of a stream at various stages of their life. It is important that a good cross section of these habitats be sampled within a given area. Specifically, try to balance your sampling units equally between pool and riffles. Document your sampling areas on your stream reach map.

2. Designate a starting point and ending point for each sample unit. The ending point should rest against a cutbank, pool tailout, or anything that can effectively trap fish in the area.

3. Measure the length, depth, and average width of each unit:

   A. Measure the length of the unit from the upstream midpoint to the downstream midpoint. "Midpoint" refers to the midpoint of the current wetted width. It is most likely not a straight line from the upstream end to the downstream end of the sample unit.

   B. Measure at least three widths across the current wetted width of the sample unit and average.

   C. At each of the three places where you measured width, measure five to 10 depths (to the nearest tenth of a foot) across the current wetted width of the sample unit and average.

4. Record your measurements. Also note if the sample unit is a riffle or a pool. Try to balance the sample units between the two types.

5. Starting at the upstream end of the unit, walk slowly downstream, moving the seine through the water. Continue moving downstream through the section. Do not raise the net until close to a containment area such as an end point location. Place netted

fish in a bucket of fresh water to be counted, measured, and identified when each unit is completed. Record the information in the appropriate sections of the data sheet.

6. Repeat the above procedure for each sampling unit in the survey section.

Hand Seining Sample Analysis

There are two methods for evaluating the hand seining results:

I. Calculate the number of fish per cubic foot:

   A. Calculate the area (cubic feet) for each unit in the sampling. Make sure all measurements are in the same measuring units before multiplying.

   $$\text{Area (cubic feet)} = L \times W \times D$$
   $$L = \text{length (ft)}$$
   $$W = \text{width (ft)}$$
   $$D = \text{depth (ft)}$$

   B. Determine the size of the total sampling area by adding the areas of all sampling units.

   C. Divide the total number of fish captured by the total sampling area to obtain the number of fish per cubic foot. This figure can be analyzed with a biologist and compared to other streams with a known population density. This evaluation will indicate whether the stream is producing up to its potential.

II. Calculate the number of fish caught per unit effort (CPUE):

   CPUE = total # fish / total # units
   Example: CPUE = 25 fish / 13 sampling units
   CPUE = 1.9 fish

Making a Hand Seine

You can make a hand seine by attaching nylon window screening to two wooden dowels. A staple gun works well, or the screening can be sewn onto the dowels with heavy duty thread. The Oregon Department of Fish and Wildlife recommends that volunteers use a seine that is four square feet in size.

The following information is reprinted with permission from the Tennessee Valley Authority, from "Homemade Sampling Equipment," Water Quality Series Booklet 2.

KICK NET

for collecting aquatic invertebrates and fish

MATERIALS:

3-foot by 6-foot piece of nylon screening
4 strips of heavy canvas (6 inches x 36 inches)
2 broom handles or wooden dowels (6 feet long)
finishing nails
thread
sewing machine
hammer
iron and ironing board

DIRECTIONS:

1. Fold nylon screen in half (3 feet x 3 feet).
2. Fold edges of canvas strips under, 1/2 inch, and press with iron.
3. Sew 2 strips at top and bottom and then use the other 2 strips to make casings for broom handles or dowels on left and right sides. Sew bottom of casings shut.
4. Insert broom handles or dowels into casings and nail into place with finishing nails.

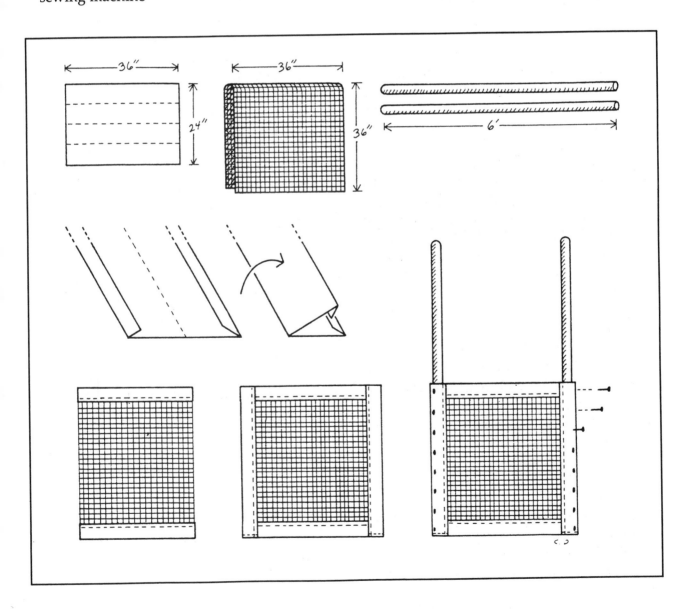

D-NET

For collecting aquatic invertebrates and fish

MATERIALS

4 pieces of nylon netting (10-inch x 12 inch)
1-inch bias tape or equivalent fabric scrap (40 inches long)
thread
scissors
sewing machine
wire coat hanger
wire cutters
drill with 1/3-inch wood bit
broom handle or wooden dowel (4 foot long)
pliers
duct tape

DIRECTIONS

1. Cut netting into four triangular pieces (10 inches high with 12-inch bases) as shown and sew together.
2. Cut a 40-inch strip of bias tape or fabric to make casing and sew onto net opening, leaving casing open to insert wire frame.
3. Take a wire coat hanger and untwist, slip into net casing, and retwist. Cut stem to 2 inches with wire cutters.
4. Drill hole in a broom handle or dowel and insert the stem as shown.
5. Take one of the remaining pieces of coat hanger and cut and bend it into a U-shape as shown.
6. Drill two shallow holes in handle, put U-shaped piece into position, push into holes as shown, and wrap with tape to secure handle.

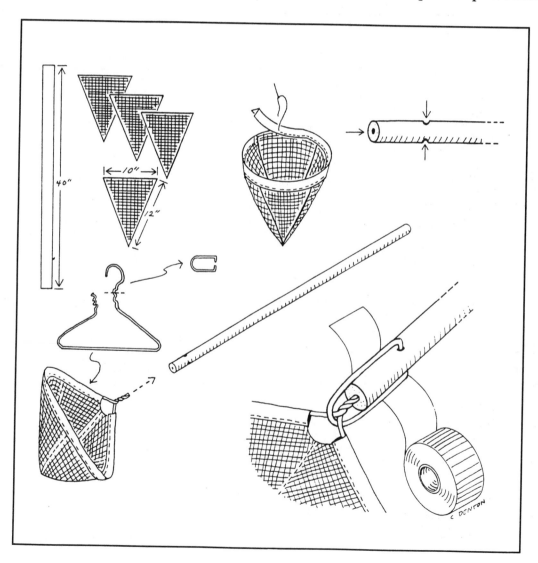

SURBER SAMPLER

For collecting invertebrates in waters less than 1 foot deep

MATERIALS:

8 straight 1-foot wood or aluminum pieces (approximately 3/16 inch x 1 inch x 1 foot in length)

2 collapsible right angle corner braces (4-6 inches long)

36 nuts and bolts (1/4 inch in diameter and 1/2 inch in length)

8 right angle braces (2 inches)

2 brass hinges (1-2 inches wide)

12 brass nutsand bolts for hinges

4 pieces of nylon netting (13 inches x 23 1/2 inches)

4 inch by 50 inch piece of heavy canvas (several pieces can be sewn together to make 50 inch strip)

bias tape

hacksaw or wood saw

screwdriver

scissors

drill with 1.4 inch and 3/16 inch bits

sewing machine

thread

adjustable wrench

ruler

pencil

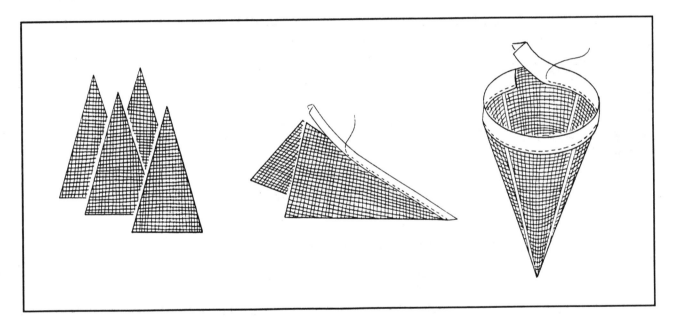

DIRECTIONS:

1. Make two squares. Use four 1-foot pieces of wood or aluminum to make each square. Mark bolt positions using ruler and pencil. Drill holes for bolts to go through. Use right angle braces to put two frames together and use wrench and screwdriver to tighten down the nuts. Leave corner of one frame unbolted to slip the net on.

2. Cut netting into 4 triangular pieces (23 1/2 inches high with 13 inch bases.)

3. Sew edges of 4 pieces of netting together using bias tape as shown.

4. To make net casing, sew 4-inch ends of canvas together to form wide cylinder. Fold in half and sew the edges of the casing to netting leaving an opening in casing to slip it onto the frame. Finished net should be 26 inches long.

5. Slip net on the unbolted corner of frame, put right angle brace in place, and tighten down nuts.

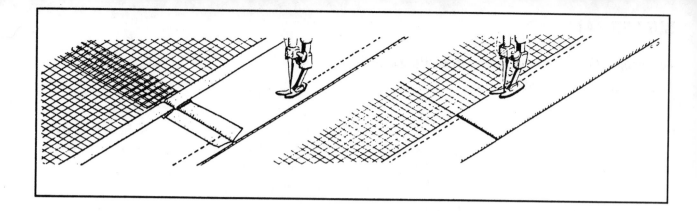

6. Lay two frames beside one another and position two hinges. Use a pencil to mark where you are going to drill. Drill holes and attach hinges. Make sure two frames fold flat.

7. Open frames to a right angle and positon collapsible right angle braces. Mark where you are going to drill with pencil, drill holes, and attach.

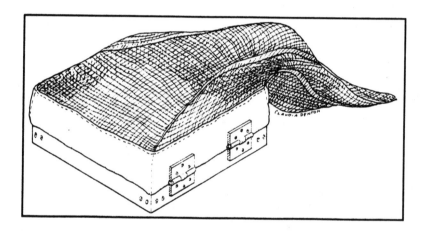

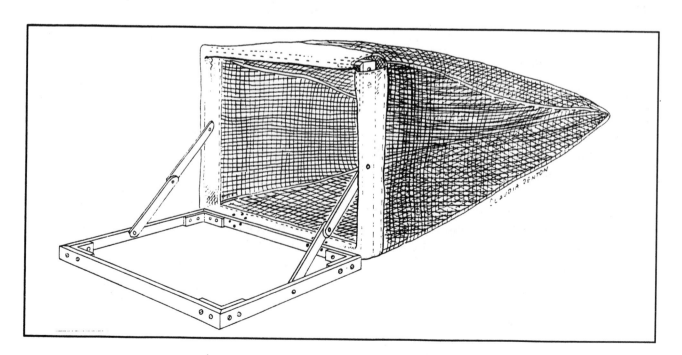

DIP NET

For collecting aquatic invertebrates and fish

MATERIALS:
wire kitchen strainer (any size) with
  handle
broom handle or wooden dowel (4
  feet long)
electrical or duct tape
scissors

DIRECTIONS
Take wire kitchen strainer and
  mount on broom handle using
  tape as shown.

COLLAPSIBLE SIEVE

For cleaning and separating organisms in an aquatic field collection

MATERIALS:
2 wire coat hangers
soldering gun and solder
1/4 -inch mesh wire screen (13-inch x 13-
  inch)
piece of heavy muslin (38-inch by 10-inch)
sewing machine
thread
needle
tin snips or kitchen scissors

DIRECTIONS:
1. Using tin snips or scissors, cut wire screen
   piece to make 13-inch diameter circle.
2. Sew 10-inch ends of muslin together to
   make wide cylinder and make 1/2-inch
   casing at top and bottom, leaving
   openings to insert wire.
3. Cut two 41-inch pieces of wire hanger,
   bend into a circle, slip into casings, solder
   hanger ends together, and hand-sew
   openings shut.
4. Put wire mesh circle in place as shown
   and hand-sew in place.

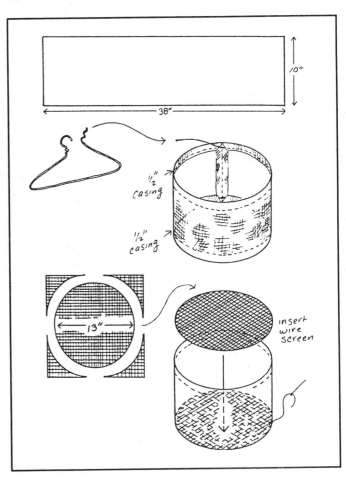

DATA SHEET 1 WATERSHED INVENTORY (PAGE 1 OF 4)

Date _____ Name_____ Group_____

Watershed Name_____ USGS quad(s)_____

Begins in_____ Flows Through_____

Ends in_____ (name towns, counties, states, regions, etc.)

Drains into_____ (name body of water)

Square Miles_____ Approx. Length_____ Width_____

Driving/Hiking Directions_____

CLIMATE

Average yearly precipitation   < 10 in ☐   10-50 in ☐   50-100 in ☐   > 100 in ☐

Most of the precipitation is in the form of   rain ☐   snow ☐   other ☐ _____

Most precipitation occurs in (month(s))_____ Floods most commonly occur in (month(s)) _____

Droughts most commonly occur in (month(s), year(s))_____

Coldest month of year_____ Warmest month of year_____ Yearly temp. range _____

GEOLOGY / TOPOGRAPHY

Describe briefly the geologic history that shaped your watershed _____

_____

(add separate sheets as necessary)

Describe the physical characteristics of different reaches of your watershed:

	Upper reaches	Middle Reaches	Lower Reaches
Uplands (mountains, hills, or flat)	_____	_____	_____
Valley (broad, medium, narrow)	_____	_____	_____
Gradient (steep, medium, gentle)	_____	_____	_____
Channel (straight, meandering)	_____	_____	_____
Bottom (boulder, cobble, gravel, fines)	_____	_____	_____

Predominate rock types   igneous ☐   sedimentary ☐   metamorphic ☐

Specific rock types present_____

Highest point_____ Lowest point: _____
(include elevation and location)

Geologic Activity: earthquakes ☐   volcanic eruptions ☐   landslides ☐   other ☐ _____

WATER RESOURCES

Headwaters originate from   glaciers ☐   snowmelt ☐   rain ☐   wetlands ☐   lakes ☐

Length of your stream _____   groundwater ☐   springs ☐

Names of tributaries_____

Order of your stream at the outlet point of your watershed _____

Names of lakes _____

Number of wetlands   few (1-15) ☐   many (15-25) ☐   Abundant (>25) ☐

Areas underlain by aquifers (if any) _____

SOILS

Predominate soil types_____

Areas with soil suitable for farming_____

Areas with soil unsuitable for development_____

Areas with potential soil erosion problems_____

VEGETATION

List the native and introduced plant species that dominate the different plant communities of your watershed:

	Native	Introduced
Upland forest	_____	_____
Riparian	_____	_____
Grassland	_____	_____
Other plant communities	_____	_____

Percent of your watershed now covered by native vegetation _____%

Reasons for the loss of native vegetation _____

Time period over which the loss occurred_____

Endangered or threatened plant species _____

FISH

Native species (circle if endangered or threatened)_____

Non-native species_____

Locations of fish hatcheries and species produced_____

Types and locations of barriers to fish migration_____

May be reproduced without permission The Adopt-A-Stream Foundation

WILDLIFE

Native species (circle if endangered or threatened) _____

_____

Non-native species _____

Key wildlife habitat areas _____

HISTORICAL (Attach extra pages as needed)

The earliest human inhabitants were _____

Describe briefly the settlement of your watershed _____

_____

Cultural and historical resources in your watershed _____

_____

DEMOGRAPHICS

Current watershed population _____ Projected population in 10 years _____ 20 years_____

Watershed population 10 years ago _____ 50 years ago _____ 100 years ago _____

Areas where most of the people live _____

List towns, cities, & counties and the percent of watershed land area that each encompasses

_____

(attach list of elected officials for each jurisdiction)

What makes people want to live (or not) in your watershed ? _____

LAND & WATER USES

Estimate the percent of your watershed zoned for each land use and check the activities that apply

rural residential _____%   densities (# of houses per acre)_____

urban/suburban residential _____%   densities _____

commercial _____% light commercial ☐   heavy commercial ☐

industrial _____%   light industry ☐   heavy industry ☐

agricultural _____% grazing ☐ confined animal ☐ dry crops ☐ irrigated crops ☐ nursery ☐

forestry _____%   clear-cut ☐   selective cut ☐   roads ☐   tree farm ☐

mining _____%   type of mining: _____

May be reproduced without permission The Adopt-A-Stream Foundation

parks/open space _____% swimming ☐   boating ☐   fishing ☐   other ☐ _____

other recreation _____%   ski resort ☐   golf course ☐   other ☐ _____

Percent of watershed that is   public land _____%   private land _____%

Percent of watershed covered with impervious surfaces _____%

Sources of domestic water supply for watershed residents _____

Location of sewage treatment plants (if any) servicing watershed residents _____

Areas that rely on septic tanks _____

Altered hydrology (dams, diversions, detention systems, culverts, dikes, drained wetlands, etc.)

type of alteration	location	purpose
_____	_____	_____
_____	_____	_____
_____	_____	_____
_____	_____	_____

(add lists as necessary)

WATER QUALITY / QUANTITY CONCERNS

Water quality classification(s) of your stream and tributaries _____

   (attach standards and regulations protecting watershed resources, include enforcing agencies)

List pollutants of concern and their potential sources (include locations if possible)

pollutant	point source	nonpoint source
_____	_____	_____
_____	_____	_____
_____	_____	_____
_____	_____	_____

AREAS PRONE TO FLOODING/DRYING UP (Add extra pages if necessary)

Location	Circle One	Dates
_____	Dry Flood	_____
_____	Dry Flood	_____
_____	Dry Flood	_____

May be reproduced without permission The Adopt-A-Stream Foundation

Name_____ Group_____ Date_____ Time_____

Stream Name_____ Reach Name/#_____

Section_____Township_____Range_____ Reach length_____

Reach Begins_____Ends_____ (in latitude/longitude or river miles)

Reach Landmarks_____

Driving/Hiking Directions_____

Weather Conditions ❑ Clear ❑ Cloudy ❑ Rain ❑ Other _____

Air temperature _____°___(C or F) Recent weather trends_____

Fish:

Type/Species (if known)	# Adults	# Juveniles	# Dead	# Redds or nests	Description & comments

Wildlife:

Birds

type, species or track/sign	# or comments

Herps (reptiles & amphibians)

type, species or track/sign	# or comments

Mammals

type, species or track/sign	# or comments

Vegetation:

Type	Abundant	Moderate	Sparse	% of 500 ft² area covered	Species present
Conifer trees					
Deciduous trees					
Shrubs					
Herbaceous					
Grasses					

	0-50'	50-100'	100+

Width of Riparian Zone: Looking downstream: Left Bank ❏   ❏   ❏
                                                    Right Bank ❏   ❏   ❏

Overhead Canopy: (at least 3' above water) 0-25% ❏   25-50% ❏   50-75% ❏   75-100% ❏

Channel Characteristics:

Gradient   Low ❏   Moderate ❏   Steep ❏   _____%
Sinuosity   Straight ❏   Meandering ❏   Braided ❏   Ranking_____

Cross section shape:

Stream Banks:

Vegetation cover: Abundant ❏   Moderate ❏   Sparse ❏   ____%

Bank stability: Erosion in some areas ❏   Erosion in many areas ❏   Intact ❏
                    Collapsed in some areas ❏   Collapsed in many areas ❏

Artificial Protection: none ❏   < 25% ❏   25-50% ❏   > 50% ❏

Describe and evaluate_____

Bank steepness: (What percent of the total length is represented by each?)

<45°___%   >45°___%   90°___%   undercut > 90___%

Reach Habitat:

or length of pool_____ divided by # or length of riffle_____ = pool : riffle ratio _____

Large Woody debris:	Abundant ❏	Moderate ❏	Sparse ❏	None ❏
Small Organic debris:	Abundant ❏	Moderate ❏	Sparse ❏	None ❏
Overhanging debris:	Abundant ❏	Moderate ❏	Sparse ❏	None ❏
Overhanging banks:	Abundant ❏	Moderate ❏	Sparse ❏	None ❏
Overhanging vegetation: (< 3 ft. above water)	Abundant ❏	Moderate ❏	Sparse ❏	None ❏
Aquatic Vegetation:	Abundant ❏	Moderate ❏	Sparse ❏	None ❏
Boulders:	Abundant ❏	Moderate ❏	Sparse ❏	None ❏

Human Alterations:

Dredging	❏	Garbage/litter	❏	Culverts	❏
Channelization	❏	Toxic substances	❏	Pipes	❏
Diversions	❏	Sewage	❏	Detention ponds	❏
Dams	❏	Bridges	❏	Storm drains	❏
Weirs	❏	Roads	❏	Other_____	❏
Dikes	❏	Other_____	❏	Other_____	❏

Land Uses:

(Enter 1 if present, 2 if you think the land use is impacting the stream.)

residential	_____	forestry	_____	grazing	_____
commercial	_____	mining	_____	crops	_____
industrial	_____	recreation	_____	irrigation	_____

Comments On Stream Reach:

DATA SHEET 3: CROSS SECTION SURVEY

Name _____ Group _____ Date _____ Time _____

Stream Name _____ Reach Name/# _____

Weather Conditions: ☐ Clear ☐ Cloudy ☐ Rain ☐ Other _____

Air Temperature: _____ ° Recent weather trends: _____

Site Name/# _____ Site Description: habitat type, landmarks, etc. _____

Site Location: Latitude _____ ° _____ ' _____ "N Longitude _____ ° _____ ' _____ "W River mile: _____

Driving/Hiking Directions _____

Depths at 1 ft Intervals:

distance (1 ft. intervals)	1	2	3	4	5	6	7	8	9	10	11	12	13	14	15	16	17	18	19	20
bankfull depth (ft)																				
wetted stream depth (ft)																				

Horizontal Distance:

Point	Left bank start point	Left bank wetted edge	Right bank wetted edge	Right bank start point
Distance	0 ft			

May be reproduced without permission The Adopt-A-Stream Foundation

A.267

Name_____ Group_____ Date_____ Time_____

Stream Name_____ Reach Name/#_____

Weather Conditions: ❏ Clear ❏ Cloudy ❏ Rain ❏ Other _____

Air Temperature: _____° ___ Recent trends: _____

Site Name/#_____

Substrate Size
(percent stream bottom that each size represents)

silt, clay, mud	
sand < 0.1"	
gravel 0.1-2"	
cobble 2-10 "	
boulder > 10"	
solid bedrock	

Site Description (habitat type, landmarks etc.)_____

_____

Site location (lat & long, transect #, and/or river mile)

_____

Embeddedness

(an estimate of the average percent that cobbles are embedded)

%

Consolidation

❏ Loose

❏ Moderately difficult to move

❏ Tightly cemented

Site Name/#_____

Substrate Size
(percent stream bottom that represents)

each size

silt, clay, mud	
sand < 0.1"	
gravel 0.1-2"	
cobble 2-10 "	
boulder > 10"	
solid bedrock	

Site Description (habitat type, landmarks etc)._____

_____

Site location (lat & long, transect #, and/or river mile)

_____

Embeddedness

(an estimate of the average percent that cobbles are embedded)

%

Consolidation

❏ Loose

❏ Moderately difficult to move

❏ Tightly cemented

Site Name/#_____

Substrate size

(percent stream bottom that each size represents)

silt, clay, mud	
sand < 0.1"	
gravel 0.1-2"	
cobble 2-10 "	
boulder > 10"	
solid bedrock	

Site Description (habitat type, landmarks etc.)_____

_____

Site location (lat & long, transect #, and/or river mile)

_____

Embeddedness

(an estimate of the average percent that cobbles are embedded)

%

Consolidation

❏ Loose

❏ Moderately difficult to move

❏ Tightly cemented

Site Name/#_____

Substrate Size

(percent stream bottom that each size represents)

silt, clay, mud	
sand < 0.1"	
gravel 0.1-2"	
cobble 2-10 "	
boulder > 10"	
solid bedrock	

Site Description (habitat type, landmarks etc.)

_____

Site location (lat & long, transect #, and/or river mile)

_____

Embeddedness

(an estimate of the average percent that cobbles are embedded)

%

Consolidation

❏ Loose

❏ Moderately difficult to move

❏ Tightly cemented

Comments

(for streams greater than 3 feet wide)

Name_____ Group_____ Date_____ Time_____

Stream Name_____ Reach Name/#_____ Site Name_____

Site Location: Latitude___ ° ___ ' ___ " N   River mile:_____
Longitude___ ° ___ ' ___ " W   Transect#:_____

Driving/Hiking directions:_____

Site Description: (e.g. transect #, landmarks, etc.)_____

Weather Conditions: ❑ Clear ❑ Cloudy ❑ Rain ❑ Other _____
Air Temperature: _____° ___ (F or C)
Amount of precipitation in last storm_____ Time elapsed since last storm_____
Other recent weather information_____

Velocity Float Trials:

Trial #	Time	Distance
1		
2		
3		
4		
5		
6		
7		
8		
9		
10		
Total		

[ ] / [ ] = [ ]
total time/# of trials   avg. time

Cross Sectional Area:

Record depths at one-foot intervals Depth = D

Cross Section 1

#	D	#	D
1		11	
2		12	
3		13	
4		14	
5		15	
6		16	
7		17	
8		18	
9		19	
10		20	

Cross Section 2

#	D	#	D
1		11	
2		12	
3		13	
4		14	
5		15	
6		16	
7		17	
8		18	
9		19	
10		20	

Cross Section 3

#	D	#	D
1		11	
2		12	
3		13	
4		14	
5		15	
6		16	
7		17	
8		18	
9		19	
10		20	

([ ] + [ ] + [ ])/3 =
  sum   sum   sum

(Multiply sums by 2 if depths were measured in 2 ft. intervals)

Average Cross Sectional Area = [ ] ft²

Average Surface Velocity = [ ] / [ ] = [ ]
distance/avg. time   feet/sec.

X (0.8) = **Average Corrected Velocity** = [ ]
velocity correction factor

Flow = [ ] ft./sec. **X** [ ] ft.² = [ ] **CFS**
Avg. Corrected Velocity   Avg. Cross Sectional Area   (cubic feet/second)

Name_____ Group_____ Date_____ Time_____

Stream Name_____ Reach Name/#_____ Site Name_____

Site Location: Latitude___°___'___" N   River mile:_____
                Longitude___°___'___" W   Transect #:_____

Site Description: (e.g. transect #, landmarks, etc.)_____
Driving/Hiking Directions_____

Weather Conditions: ☐ Clear ☐ Cloudy ☐ Rain ☐ Other _____
                    Air Temperature: _____°___ (C or F)
Amount of precipitation in last storm_____ Time elapsed since last storm_____
Other recent weather information_____

Velocity Float Trials:

trial #	time	Distance
1		
2		
3		
4		
5		
6		
7		
8		
9		
10		

[ ] / [ ] = [ ]
sum/# of trials   avg time

Average Surface Velocity = [/] = _____
distance/avg. time   feet/sec.

Average Corrected Velocity = _____ X (0.8) = [ ]
                          feet/sec   velocity correction   feet/sec
                                     factor

Use Average Corrected Velocity to calculate
Flow, using formula on bottom of next page.

Cross Section 1

1	
2	
3	

avg. depth = ☐ **/3** = ☐
           sum   feet

$Area_1 =$ ☐ **x** ☐ = ☐ ft^2
     avg. depth   width

Cross Section 2

1	
2	
3	

avg. depth = ☐ **/3** = ☐
           sum   feet

$Area_2 =$ ☐ **x** ☐ = ☐ ft^2
     avg. depth   width

Cross Section 3

1	
2	
3	

avg. depth = ☐ **/3** = ☐
           sum   feet

$Area_3 =$ ☐ **x** ☐ = ☐ ft^2
     avg. depth   width

Use Average Cross Sectional Area to calculate Flow, using formula on bottom of this page.

Average Cross Sectional Area =

☐
$Area_1$

+

☐
$Area_2$

+

☐ **/3**
$Area_3$

= ☐ $feet^2$

Flow = ☐ **X** ☐ = ☐

      Avg. Corrected   Avg.Cross   **CFS**
        Velocity   Sectional Area   (cubic feet/second)

DATA SHEET 6A MACROINVERTEBRATE SURVEY (FIELD METHOD)

Name_____ Date_____ Time_____

Stream Name_____ Reach Name/#_____ Site Name/#_____

Driving/Hiking Directions_____

Weather Conditons: ____Clear   ____Cloudy   ____Rain   ____Other_____

Air Temperature_____°__   Water Temperature_____°__

Recent Weather Trends_____

major group	# taxa in each Feeding Group* SH	SC	GC	FC	P	# taxa
mayflies						
stoneflies						
caddisflies						
true flies: midges						
craneflies						
blackflies						
others						
dobsonflies,alderflies, fishflies						
beetles						
dragonflies/damselflies						
crustaceans: crayfish						
sowbugs						
scuds						
others						
snails						
clams & mussels						
worms & leeches						
others						
Totals ➝						

Taxa in each functional feeding group

Taxa Richness (total # of taxa)

Total # mayfly, stonefly, caddisfly taxa

EPT Richness

Site Description:
Habitat Type_____
_____

Avg. depth_____

Avg. current velocity
_____

Other Comments_____
_____
_____

Site Location (lat/long. transect # and/or

rivermile_____

*Major Feeding Groups

SH - Shredder
SC - Scraper
GC - Gatherer Collector
FC - Filter Collector
P - Predator

Name_____ Group_____ Date_____ Time_____
Stream Name_____ Reach Name/#_____ Site Name_____
Site Description: Habitat type_____ Avg. depth_____
Avg. current velocity_____ Other comments_____
Site Location (lat. & long., transect #, and/or river mile_____
Driving/Hiking Directions_____
_____

Weather Conditions: ☐ Clear ☐ Cloudy ☐ Rain ☐ Other_____
Air Temp. ____ Water Temp. ____ Recent weather trends:_____

Families in major groups	feeding group	density (D)			avg. D	tolerance value	tolerance X avg. D	avg. D (major group)
		rep.1	rep.2	rep.3				
mayflies							→	
stoneflies							→	
caddisflies							→	
total avg. density EPT (mayfly, stonefly, caddisfly) →								

Families in major groups	tolerance value	density (D)			avg. D	feeding group	tolerance X avg. D	avg. D (major group)
		rep.1	rep.2	rep.3				
midges (fam. chironomidae only)								
other true flies							→	
dobsonflies, alderflies, fishflies							→	
beetles							→	
dragonflies, damselflies							→	
crustaceans							→	
snails							→	
clams, mussels							→	
worms, leeches							→	
others								
total tolerance x D							→	

total average density of sample

Data Summary + Analysis

1. Taxa Richness (total # taxa)   =   [　　　]

2. EPT Richness (total mayfly, stonefly, caddisfly taxa)   =   [　　　]

3. Projected Average Density for the sample =   average of the projected densities for each replicate:

replicate #   (total # bugs picked/total # squares picked) x 12

replicate #			
1	(/) x 12 =		
2	(/) x 12 =		
3	(/) x 12 =		

average = [　　　]

4. Percent Composition of Major Groups $\left(\dfrac{\text{total avg. density for each major group}}{\text{total avg. density of sample}}\right)$

[　] mayflies	[　] crustaceans
[　] stoneflies	[　] snails
[　] caddisflies	[　] clams, mussels
[　] midges	[　] worms, leeches
[　] other true flies	[　] others
[　] dobsonflies, alderflies, etc.	
[　] beetles	

5. Percent Composition of Functional Feeding Groups

	Shredders	Scrapers	Gathering Collectors	Filtering Collectors	Predators
total = sum avg. # density for each					
%= tot / tot. avg. density of sample					

6. EPT/Midge Ratio $\dfrac{\text{Total. Avg. density EPT}}{\text{Total Avg. density midges}}$ =   [　　　]

7. Family Biotic Index $= \dfrac{\text{the sum of all "tolerance X D" Columns}}{\text{Total average density of sample}}$ =   [　　　]

DATA SHEET 7 WATER QUALITY

Name_____ Group_____ Date_____ Time_____

Stream Name_____ Reach Name/#_____

Site Description: (site name, landmarks, etc.)_____

Site Location: (lat & long, transect #, and/or rivermile) _____

Driving and/or Hiking Directions_____

Weather Conditions: ❑ Clear ❑ Cloudy ❑ Rain ❑ Other _____

Air temperature: _____° ___ Recent weather trends:_____

Quantitative Water Quality

pH

1_____

2_____

3_____

mean_____

Dissolved O_2

1_____

2_____

3_____

mean _____

Temperature

_____ ° ___

_____% Dissolved
Oxygen
Saturation

Qualitative Water Quality

Water Appearance
❑ Scum
❑ Foam
❑ Muddy
❑ Milky
❑ Clear
❑ Oily sheen
❑ Brownish
Other_____

Stream Bed Coating
❑ Orange to red
❑ Yellowish
❑ Black
❑ Brown
❑ None

Other_____

Odor
❑ Rotten egg
❑ Musky
❑ Acrid
❑ Chlorine
❑ None

Other_____

Other Tests:

Comments:

APPENDIX D. EQUIPMENT SOURCES

SOURCES	Stream surveying equipment	Water testing kits	Water testing meters	Thermometers	Flow measuring equipment	Plant collecting equipment	Bug collecting equipment	Fish collecting equipment	Portable microscopes	Waders and rubber boots	Lab items	Educational items		
Archie McPhee Box 30852 Seattle, WA 98103 (206) 547-2467					*				*					
Ben Meadows Company P.O. Box 80549 Chamblee, GA 30366 800 241-6401 FAX (404) 457-1841	*			*	*					*				
BioQuip Products 17803 LaSalle Ave. Gardena, CA 90248 (213) 324-0620							*		*		*	*		
Cabela's 812 13th Ave Sydney, NA 69160 800 237-4444 FAX (308) 254-2200										*				
Carolina Biological Supply Company Powell Laboratories 19355 McLoughlin Boulevard Gladstone, OR 97027 800 547-1733 FAX (503) 656-4208				*		*	*	*	*		*	*		
Main Office and Laboratories 2700 York Road Burlington, NC 27215 800 334-5551 FAX (919) 584-3399				*		*	*	*	*		*	*		
ConAgra Technologies P.O. Box 128, Route 117 N. Goodfield, IL 61742 800 634-7571		*												

EQUIPMENT SOURCES

SOURCES	Stream surveying equipment	Water testing kits	Water testing meters	Thermometers	Flow measuring equipment	Plant collecting equipment	Bug collecting equipment	Fish collecting equipment	Portable microscopes	Waders and rubber boots	Lab items	Educational items		
Edmund Scientific 101 E. Gloucester Pike Barrington, NJ 08007-1380 (609) 573-6260 FAX (609) 573-6295			*	*					*		*	*		
Forest Densiometers 2413 N Kenmore Street Arlington, VA 22207 (703) 528-2851	*													
Forestry Suppliers Inc. 205 W. Rankin St. P.O. Box 8397 Jackson, MS 39284-8397 800 647-5368 FAX 800 543 4203	*			*	*	*				*	*			
Hach Company P. O. Box 389 Loveland, CO 80539 800-227-4224 FAX (303) 669-2932		*	*									*		
Idaho Nets and Benthic Supply Teton, ID 83451 (208) 458-4594							*	*		*				
J.L. Darling Company (All Weather Writing Paper) 212 Port of Tacoma Road Tacoma, WA 98421 (206) 383-1714 FAX (206) 383-1722	*													
Kahl Scientific Instrument Corporation P.O. Box 1166 El Cajon, CA 92022-1166 (619) 444-2158	*	*	*	*	*	*	*	*			*			

EQUIPMENT SOURCES

SOURCES	Stream surveying equipment	Water testing kits	Water testing meters	Thermometers	Flow measuring equipment	Plant collecting equipment	Bug collecting equipment	Fish collecting equipment	Portable microscopes	Waders and rubber boots	Lab items	Educational items		
LaMotte Chemical Products Company P.O. Box 329 Chestertown, MD 21620 800-344-3100 FAX (301) 778 6394		*									*			
Museum Products 84 Route 27 Mystic CT 06355 800-395-5400									*			*		
Research Nets, Inc. P.O. Box 249 Bothell, WA 98041 (206) 821-7345 FAX (206) 775-5122							*	*						
Scientific Instruments, Inc. 518 W. Cherry St. Milwaukee, WI 53212 (414) 263-1600	*				*									
Solomat Instrumentation Division Glenbrook Industrial Park 652 Glenbrook Road Stamford, CT 06906 (203) 348-9700	*		*		*									
Swoffer Instruments 1048 Industry Drive Seattle, WA 98188 (206) 575-0160 FAX (206) 575-1329	*				*									
Thomas Scientific 800 345-210				*										

SOURCES	Stream surveying equipment	Water testing kits	Water testing meters	Thermometers	Flow measuring equipment	Plant collecting equipment	Bug collecting equipment	Fish collecting equipment	Portable microscopes	Waders and rubber boots	Lab items	Educational items
Ward's Biological P.O. Box 5010 San Luis Obispo, CA 93403 800 962-2660				*		*	*	*	*		*	*
Wildlife Supply Company 301 Cass Street Saginaw, MI 48602 (517) 799-8100 FAX (517) 799-8115	*		*	*	*	*	*	*			*	
Yellow Springs Instruments P.O. Box 279 Yellow Springs, OH 45387 (513) 767-7241	*		*	*							*	

The following is an element of the *Quality Assurance Project Plan for the Citizen Monitoring Project*, Citizen's Program for the Chesapeake Bay, Inc.

EQUIPMENT NEEDED:

1 precision grade thermometer that reads in 0.1 increments, degrees C.
1 insulated cooler
1 wide-mouthed large jar, e.g., 1 quart mayonnaise jar
String or twine
ice

PROCEDURE:

1. Fill the insulated cooler with tap water and let it equilibrate inside overnight with a precision thermometer in the water.
2. The following day:
   a. With the string, tie the field thermometers to be calibrated loosely together. Suspend in water in cooler. Do no more than 10 thermometers at a time. Let stand for 15 minutes. Read the precision thermometer and record value. Read all the field thermometers and record values. Let stand another 15 minutes and take a second reading on all thermometers.
   b. Prepare an ice bath in the wide-mouthed jar. Make sure that the ice to water ratioo is such that the jar is packed with ice to the bottom. Place precision and field thermometers in ice bath. If possible, suspend in the center of the bath by tying string to a cabinet door, (or any other higher object being sure that the arrangement is not going to fall over!) Let stand 15 minutes. Take and record readings. Wait another 15 minutes, adding more ice if remaining ice is floating above bottom of jar. Take second readings.
   c. Analyze Results. The thermometers should read to within 1.0 degree C of the precision thermometer. Any thermometer outside this range should not be used in your program. (If in doubt, repeat calibration procedures. If still out of range, do not use.)

Editors Note: Assign each of your thermometers an inventory number (list in your QA plan), and record any temperature variations detected that are within 1.0 degree C of the precision thermometer on your inventory sheet.

APPENDIX F CONVERSIONS AND EQUIVALENTS

Length

1 mm	= 0.1 cm = 0.001 m = 0.0394 in	**in**	= 25.4 mm = 2.54 cm	**1 m**	= 100 cm = 3.281 ft =39.37 in	**1 yd**	= 36 in = 91.44 cm = 0.9144 m
1 cm	= 0.01m = 0.394 in = 0.0328 ft	**1 ft**	= 12 in = 30.48 cm = 0.3048 m	**1 km**	= 100,000 cm = 1000 cm = 0.6214 mi	**1 mi**	= 5280 ft = 1609 m = 1.609 km

Area

1 mm^2	= 0.01 cm^2 = 0.00155 in^2	**1 in^2**	= 6.452 cm^2	**1 ha**	= 10,000 m^2 = 2.471 ac = 0.00386 mi^2	**1 ac**	= 43560 ft^2 = 4840 yd^2 = 4047 m^2 = 0.4047 ha
1 cm^2	= 100 mm^2 = 0.1550 in^2	**1 ft^2**	= 144 in^2 = 929.0 cm^2	**1 km^2**	= 247.1 ac = 0.3861 mi^2	**1 mi^2**	= 27,878,400 ft^2 = 640 ac = 259 ha = 2.59 km^2
1 m^2	= 10,000 cm^2 = 1550 in^2 = 10.76 ft^2	**1 yd^2**	= 9 ft^2 = 8361 cm^2				

Volume

1 kg	= 2.205 lb	**1 in^3**	= 16.39 cm^3	**1 ml**	= 0.061 in^3
1 m^3	= 35.31 ft^3 = 264.2 US gal	**1 ft^3**	= 1728 in^3 = 7.48 US gal = 28317 cm^3 = 28.32 liters	**1 liter**	= 61.02 in^3 = 0.2642 US gal

Weight

1 kg	= 2.205 lb	**1 lb**	= 0.4536 kg
1 g	= 0.001 kg = 0.0353 oz = 0.0022 lb	**1 grn**	= 0.0648 g
		1 oz	= 28.35 g
		1 ton	= 2000 lb = 907.2 kg

Temperature

F∞	0	+10	20	30	40	50	60	70	80	90	100
C∞	-17.8	-12.2	-6.7	-1.1	+4.4	10	15.6	21.1	26.7	32.2	37.8

C∞	-10	0	+10	20	30	40
F∞	+14.0	32	50.0	68.0	86.0	104.0

$$C\infty = (5/9) \times (F - 32)$$

$$F\infty = [(9/5) \times C] + 32$$

Velocity

1 m/s	= 3.6 km/hr		1 ft/sec	= 0.6818 mi/hr
	= 2.237 mi/hr			= 1.097 km/hr
1 km/hr	= 0.2278 m/s		1 mi/hr	= 1.4767 ft/sec
	= 0.9113 ft/sec			= 0.447 m/sec

Discharge

1 m³/sec	= 35.32 cfs	1 cfs	= 0.0283 m³/sec	1 cfs for 1 day	= 1.98 acre ft
			= 28.32 l/sec		
			= 2447 m³/day		

Hydrologic Information

1 in/hr of runoff from 1 acre = 1 cfs
1 acre ft = 325,851 gal

Governments

US	Canada	Mexico
who?	who?	quien?
what?	eh?	que?
where?	where?	donde?
Federal	Federal	Federale
State	Province	Estado
County/Parish	Regional District	Municipio
City/Town	Municipality/City/ Town/Village/ Township	Ciudad/Pueblo

CURRICULUM MATERIALS

Aquatic Resources Education Curriculum. Kendall Hunt Publishing Co., P.O. Box 539. Dubuque, IA

Delta Labs.1987. *Adopt-A-Stream Teacher's Handbook.* Delta laboratories, Rochester, NY.

Dyckman, C. and A. W. Way. 1981. *Clean Water, Streams and Fish: A Holistic View of Watersheds.* Municipality of Metropolitan Seattle, Seattle, WA.

Farthing, P. Bill Hastie, Shann Weston, Don Wolf. 1992. *The Stream Scene: Watersheds, Wildlife and People.* Oregon Department of Fish and Wildlife, Portland, OR.

Gray-Hubbard, S. and S. Tilander. 1989. *Stream Team Guidebook.* City of Bellevue Storm and Surface Water Utility. City of Bellevue, Bellevue, WA.

Higgins, D. 1990. *California's Salmon and Steelhead: Our Valuable National Heritage.* California Advisory Committee on Salmon and Steelhead Trout, 1271 Fieldbrook Road, Arcata, CA 95521. (707) 822-0744.

Lewis, B. 1991. *The Kid's Guide to Social Action: How to Solve the Social Problem You Choose----and Turn Creative Thinking into Positive Action.* Free Spirit Publishing, Minneapolis MN.

Living in Water: An Aquatic Science Curriculum. National Aquarium in Baltimore, Department of Education and Interpretation, Pier 3, 501 E. Pratt St., Baltimore, MD.

Mitchell, M. K. and W. B. Stapp. 1990. *Field Manual for Water Quality Monitoring: An Environmental Education Program for Schools.* Thompson-Shore Printers, Dexter, MI.

Salmonids in the Classroom, Primary and Secondary. BCTF Lesson Aids 105-2235 Burrand St., Vancouver, B.C. Canada, V6J 3N9. (604) 731-8121.

Save Our Streams. Izaak Walton League of America. 1701 N. Fort Meyer Dr. #1100, Arlington, VA 22209.

United Nations. 1977. *Small Streams and Salmonids: A Handbook for Water Quality Studies.* United Nations Educational, Scientific and Cultural Organization, Paris, and Municipality of Metropolitan Seattle, Seattle, WA.

Water, Water Everywhere. Hatfield Marine Science Center, Extension Sea Grant, Newport, OR 97365.

Western Regional Environmental Education Library. 1987. Aquatic Project Wild.

FISH

Bond, C.E., 1973. *Key to Oregon Freshwater Fishes*, rev. ed. Oregon Agriculture Experiment Station, Technical Bulletin 58.

Brown, E. R. 1985 *Management of Wildlife and Fish Habitats in Forests of Western Oregon and Washington.* USDA Forest Service Pacific Northwest Region, Portland, OR.

Brown, Bruce. 1982. *Mountain in the Clouds: The Search for the Wild Salmon.* Simon and Schuster, NY.

Carl, G. C., W.A. Clemens, and C.C. Lindsey, 1967. *The Freshwater Fishes of British Columbia.* 4th ed. British Columbia Provincial Museum, Victoria.

Clifford, Carl; Clemens, W.A.; and Lindsey; 1977. *The Freshwater Fishes of British Columbia.* British Columbia Provincial Museum, Victoria, Canada.

Eddy, Samuel and Underhill, James. 1978. *How to Know the Freshwater Fishes.* Wm. C. Brown Company Publishers.

Hunter. C.J., 1991. *Better Trout Habitat: A guide to Stream Restoration and Management.* Montana Land Reliance. Island Press, Washington D.C.

McAllister, D.E. and E.J. Crossman, 1973. *A Guide to the Freshwater Sport Fishes of Canada.* National Museum of Natural Scieces, Ottawa, Canada.

Moyle, Peter B., 1976. *Inland Fishes of California.* University of California Press, Berkeley, CA.

Netboy, A. 1958. *Salmon of the Pacific Northwest: Fish vs. Dams.* Binfords and Mort, Portland.

Netboy, A. 1973. *The Salmon, Their Fight for Survival.* Houghton Mifflin Co., Boston, MA.

Northwest Power Planning Council. *Strategy for Salmon.* Columbia River Basin Fish and Wildlife Program. Northwest Power Planning Council.

Royce, William F. 1972. *Introduction to the Fishery Sciences.* Academic Press. New York, NY.

Scott, W. B. and E. J. Crossman. 1973. *Freshwater Fishes of Canada.* Fisheries Research Board of Canada Bulletin 184, Toronto Canada.

Wydoski, R. S. and R. R. Whitney, 1978. *Inland Fishes of Washington.* University of Washington Press, Seattle, WA.

GROUNDWATER

Freeze, R. A. and J. A. Cherry. 1979. *Groundwater.* Prentice-Hall.

Simpson, J.T. 1991. *Volunteer Lake Monitoring: A Methods Manual.* U.S. Environmental Protection Agency. Office of Water, Washington, D.C.

Todd, D. K. 1980. *Groundwater Hydrology.* Wiley Interscience.

LAKES

Freshwater Foundation. 1985. *A Citizen's Guide to Lake Protection*. Freshwater Foundation, Navarre, MN.

Moore, M. L. 1989. NALMS *Management Guide for Lakes and Reservoirs*. North American Lake Management Soiciety, Washington D.C.

MACROINVERTEBRATES

Bauman, R. W., A. R. Gaufin and R. F. Surdick. 1977. *The Stoneflies of the Rocky Mountains*. Memoirs of the American Entomological Society, Academy of Natural Sciences, Philadelphia, PA.

Cummins, K.W. and Wilzbach, M.A. 1985. *Field Procedures for Analysis of Functional Feeding Groups of Stream Macroinvertebrates*. University of Maryland, Appalachian Environmental Laboratory.

Edmondson, W. T. 1959. *Freshwater Biology*. (2nd edition). John Wiley and Sons, New York.

Edmunds, G. F., S. L. Jensen and L. Berner. 1976. *The Mayflies of North and Central America*. University of Minnesota Press, Minneapolis, MN.

Kellogg. L. L. *Save Our Streams: Monitor's Guide to Aquatic Macroinvertebrates*. Izaak Walton League, Arlington, VA.

Lehmkuhl, D. M. 1979. *How to Know Aquatic Insects*. W. C. Brown, Dubuque, IA.

McAlpine, J. F., B. V. Peterson, G. E. Shewell, H. J. Teskey, J. R. Vockeroth and D. M. Wood. (coordinators). 1981. *Manual of Nearctic Diptera*. Volumes 1 and 2. Res. Branch, Agric. Can. Monogr. 27.

McCafferty, W. P. 1981. *Aquatic Entomology*. Science Books International, Boston, MS.

Merritt, R. W. and K. W. Cummins. 1987. *An Introduction to the Aquatic Insects of North America*. Kendall-Hunt, Dubuque, IA.

Needham, J. G. and P. R. Needham. 1962. (5th edition). *A Guide to the Study of Freshwater Biology*. Holden-Day, San Francisco, CA.

Pennak, R. W. 1990. (2nd edition). *Freshwater Invertebrates of the United States*. John Wiley, New York, NY.

Stewart, K. W. and B. P. Stark. 1989. *Nymphs of the North American Stonefly Genera (Plecoptera)*.

U.S. Environmental Protection Agency. 1989. *Rapid Bioassessment Protocols for use in Streams and Rivers: Benthic Macroinvertebrates and Fish*. Office of Water, Washington, D.C.

Usinger, R. L. 1956. *Aquatic Insects of California*. University of California Press, Berkeley, CA.

Wiggens, G. B. 1977. *Larvae of the North American Caddisfly Genera (Trichoptera)*. University of Toronto Press, Toronto, Canada.

MAPS

Birkby, R. 1990. *The Boy Scout Handbook.* Boy Scouts of America, Irving, Texas.

Boy Scouts of America. 1988. *Merit Badge Pamphlet for Orienteering.* ($1.39; available from Boy Scouts of America, 1325, Walnut Hill Lane, Irving, TX, 75038.

Jacobson, C. 1988. *The Basic Elements of Map and Compass.* ICS Books, Merrillville, IN.

Kals, W.S. 1983. *Land Navigation Handbook-The Sierra Club Guide to Map and Compass.* Random House, NY.

Kjellstrom, B. 1976. *Be Expert with Map and Compass-The Orienteering Handbook.* Charles Scribner and Sons, NY.

Makower, J. (Ed.). 1990. *The Map Catalog: Every Kind of Map and Chart on Earth and Even Some Above It.* 2nd Edition. Vintage Books, NY.

Ratliffe, D. E. 1979. *Map, Compass and Campfire.* Binford and Mort, Portland, OR

USGS. 1988. *Washington index to topographic and other map coverage.* Western Mapping Center, USGS, 345 Middlefield Road, Menlo Park, CA 94025. (can be obtained from DNR's map store in Olympia or sometimes at local map or sporting goods stores.)

USGS. 1988. *Washington catalog of topographic and other published maps.* Western Mapping Center, USGS, 345 Middlefield Road, Menlo Park, CA 94025. (can be obtained from DNR's map store in Olympia or sometimes at local map or sporting goods stores.)

QUALITY ASSURANCE

Tetra Tech, Inc. 1994. *Generic Quality Assurance Project Plan Guidance for Programs Using Community Level Biological Assessment in Streams and Rivers.* Prepared for the EPA, by Tetra Tech, Inc., Owings Mills, MD.

Washington State Department of Ecology. 1991. *Guidelines and Specifications for Preparing Quality Assurance Project Plans.* Environmental Investigations and Laboratory Services Program, Quality Assurance Section, Manchester. WA.

RIPARIAN AREAS

Reichard, N. 1987. *Stream Care Guide for Streamside Property Owners and Residents.* Miller Press, Eureka, CA.

Maser, C. et al. 1988. *From the Forest to the Sea: A Story of Fallen Trees.* General Technical Report PNW-GTR-229. USDA Forest Service, Pacific Northwest Research Station, Portland, OR.

Cundy, T. and E. Salo (editors). *Streamside Management: Forestry-Fisheries Interactions.* Institute for Forest Resources, University of Washington , Seattle, WA.

Meehan, W. R. et al. 1984. *Fish and Wildlife Relationships in Old-Growth Forests.* American Institute of Fisheries Research Biologists, Juneau, AK.

Oregon Watershed Improvement Coalition. *Riparian Areas: Their Benefits and Uses: What is a Riparian Area?*

Raedeke, K. J. (ed.) 1988. *Streamside Management: Riparian Wildlife and Forestry Interactions.* Contribution 59, Institute for Forest Resources, University of Washington. Seattle, WA.

STREAM ECOLOGY

Cummins, K. W. 1974. *Structure and function of stream ecosystems.* Bioscience 631-641.

Duncan, D. J. 1983. *The River Why.* Sierra Club Books, San Francisco, CA.

Hynes, H. B. N. 1975. *Ecology of Running Water.* University of Toronto Press, Toronto, Canada.

Ministry of Environment. 1980. *Stream Enhancement Guide.* British Columbia, Canada.

Oglesby, R. T. at al. 1972. *River Ecology and Man.* Academic Press, NY.

Townsend, C. R. 1980. *The Ecology of Streams and Rivers.* Edward Arnold, London.

Water Watch. 1986. *A Field Guide to Kentucky Rivers and Streams.* Kentucky Natural Resources and Environmental Protection Cabinet, Frankfurt, KY.

Whitton, B.A. 1975. *River Ecology.* University of California Press.

Winborne, F. B. 1989. *A Guide to Streamwalking.* Stream Watch, Division of Water Resources, Raleigh, NC.

Yates, S. 1989. *Adopting A Stream: A Northwest Handbook.* University of Washington Press, Seattle, WA.

WATER QUALITY

APHA. 1989. *Standard Methods for the Examination of Water and Wastewater.* 17th Edition. American Public Health Association, Washington D.C.

Brown, George W. 1985. *Forestry and Water Quality.* 2nd edition. Oregon State University, Corvallis, OR.

Campbell, G. and Wildberger, S. 1992. *The Monitor's Handbook.* LaMotte Company, Chestertown, MD.

Cusimano, R.F. 1994. *Technical Guidance for Assessing the Quality of Aquatic Environments.* Washington State Department of Ecology. Environmental Investigations and Laboratory Services Program, Olympia, WA.

HACH Company. *Laboratory pH/ISE Meter pH Quick Reference Guide.* HACH Company World Headquarters, Loveland, CO.

Hinman, K. 1994. *"Recent News About the Hanford Reach."* Douglasia, Winter 1994.

MacDonald, L.H. with Smart, A.W. and Wissman, R.C. 1991. *Monitoring Guidelines to Evaluate Effects of Forestry Activity on Streams in the Pacific Northwest and Alaska.* U.S. Environmental Protection Agency, Region 10, Seattle, WA.

Platts, W. S. et al. 1983. *Methods for Evaluating Stream, Riparian and Biotic Conditions.* Ogden, UT: US Department of Agriculture.

Puget Sound Estuary Program. 1990. *Recommended Protocols for Measuring Conventional Water Quality Variables and Metals in Freshwater of the Puget Sound Region.* Prepared for U.S. EPA Region 10, Office of Puget Sound, Seattle, WA.

Reisner, M. 1986. *Cadillac Desert: the American West and its Disappearing Water.* Penguin Boks, NY.

Richter, J. 1988. *State of the Sound.* Puget Sound Water Quality Authority. Seattle, WA.

River Watch Network. 1994. *Chemical Water Quality Manual.* River Watch Network, Montpelier, VT.

Tennessee Valley Authority. *Water Quality Sampling Equipment: Water Quality Series Booklet 1.* Tennessee Valley Authority's Teacher/Student Water Quality Monitoring Network.

Tennessee Valley Authority. *Homemade Sampling Equipment: Water Quality Series Booklet 2.* Tennessee Valley Authority's Teacher/Student Water Quality Monitoring Network.

Terell, C. R. and P. B. Perfetti. 1989. *Water Quality Indicators Guide: Surface Waters.* U.S. Soil Conservation Service, Washington DC.

U.S. EPA. 1983. *Methods for Chemical Analysis of Water and Wastes.* EPA-600/4-79-020. U.S. EPA Environmental and Support Laboratory, Cincinnatti, OH.

U. S. EPA. 1987. *Guide to Nonpoint Source Pollution Control.* Office of Water, Washington DC.

U. S. EPA. 1990. *The Quality of Our Nation's Water.* EPA 440/4-90-005. Office of Water, Washington DC.

U. S. EPA. 1990. *Volunteer Water Monitoring: A Guide for State Managers.* EPA 440/4-90-010. Office of Water, Washington DC.

Widmer K. 1988. *Project Mayfly.* A guide to the determination of water pollution in local waterways. National Audubon Society, Camp Hill, PA.

WATERSHEDS

Coyote Creek Riparian Station. *1994. Riparian Habitat Inventory of Santa Clara County Protocols and Procedures, Progress Report,* January 1994. Santa Clara County Stream Inventory.

Dunne, T.B. and L.B. Leopold. 1978. *Water in Environmental Planning.* W.H. Freeman, San Francisco, CA.

House, F. 1990. *To learn the things we need to know: engaging the particulars of the planet's recovery.* Whole Earth Review 66: 36-47.

Hynes, H.B.N. 1975. *The stream and its valley.* Verh. Internat. Verein. Limnol. 19: 1-15.

Leopold, L. B. 1974. *Water, A Primer.* W. H. Freeman and Co., San Francisco, CA.

Likens, G.E. and F.H. Borman. 1974. *Linkages between terrestrial and aquatic systems.* Bioscience 24 (8): 447-456.

Lotspeich, F.B. 1980. *Watersheds as the basic ecosystem: this conceptual framework provides a basis for a natural classification system.* Water Resources Bulletin 16(4): 581-586.

Penman, H.L. 1970. *The water cycle in the biosphere.* Scientific American. W.H. Freeman and Company, San Francisco, CA.

U.S. Forest Service, Region 6. 1974-1975. *Stream Inventory Handbook,* Level I & II, Version 7.

Warshall, P. 1976. *Streaming Wisdom: Watershed Consciousness in the Twentieth Century.* The CoEvolution Quarterly. Volume 12.

WETLANDS

Cowardin, L. M. et al. 1979. *Classification of Wetlands and Deepwater Habitats of the United States.* Fish and Wildlife Service, U. S. Department of the Interior, Washington D.C.

Frost, J. R. 1990. *Wetlands Preservation: An Information and Action Guide.* Washington State Department of Ecology Olympia, WA.

Kusler, J. A. 1983. *Our National Wetland's Heritage: A Protection Guidebook.* Island Press, Covelo, CA. Answers frequently asked questions about wetlands and discusses practical ways to preserve and protect them.

McMillan, A. 1985. *Washington's Wetlands.* Washington State Department of Ecology, Olympia, WA.

Michaud, J. P. 1990. *At Home with Wetlands; a Landowner's Guide.* Washington State Department of Ecology, Olympia, WA.

Mitsch W. and J. Gosselink. 1986. *Wetlands.* Van Nostrand Reinhold, NY.

Niering, W.A. *The Audubon Society Nature Guides: Wetlands.* The Audubon Society. Alfred A. Knopf.

Weller, M. F. 1981. *Freshwater Marshes.* University of Minnesota Press, Minneapolis, MN.

Yates, S. 1990. *Adopting A Wetland: Northwest Guide.* Snohomish County Planning and Community Development,WA

Water Watch. 1986. *A Field Guide to Kentucky Lakes and Wetlands.* Kentucky Natural Resources and Environmental Protection Cabinet. Frankfurt, KY.

GENERAL ARTICLES

River Network. *River Voices,* Summer 1993. Portland, OR.

Index

T

U

V

W

Z

ABOUT THE AUTHORS

Tom Murdoch is the Founder and Executive Director of the Adopt-A-Stream Foundation. He has also worked for local government developing drainage regulations and regional surface water management plans, and directing the inventory and mapping of more than 5000 acres of wetlands and 300 miles of streams. During the last 17 years, Tom has become a practicing stream ecologist responsible for more than 200 stream restoration projects, numerous national and international stream and wetland restoration articles and presentations, and has assisted hundreds of volunteers to "adopt" streams. He holds a Masters degree in Public Administration from the University of Washington's School of Public Administration and Institute for Marine Studies.

Martha Cheo is a former Assistant Director of the University of Rhode Island's W. Alton Jones Environmental Education Center where she authored *The Pawcatuck Watershed Education Program Curriculum Guide* and taught stream ecology. Between 1993 and 1995, she was the Adopt-A-Stream Foundation's Ecologist responsible for organizing and conducting Adopt A Stream and Streamkeeper Field Training workshops throughout the Pacific Northwest. Martha, who has a Masters in Natural Science from the University of Rhode Island, currently works as an environmental educator for the Institute of Ecosystem Studies in New York.

Kate O'Laughlin was the Adopt-A-Stream Foundation's Ecologist during 1991 and 1992, and author of the first edition of *the Streamkeeper's Field Guide*. She is the author of numerous technical papers, an expert on benthic macroinvertebrates, and veteran instructor of Streamkeeper Field Training workshops. Kate, who has a Masters of Science Degree in Biology from Eastern Washington University, is currently a staff ecologist for the King County (WA) Surface Water Management Department's stream monitoring program.

ALSO AVAILABLE FROM THE ADOPT-A-STREAM FOUNDATION

The Streamkeeper
starring Bill Nye "The Science Guy"

A VALUABLE COMPANION to *the Streamkeeper's Field Guide, The Streamkeeper* video is an upbeat training tool for anyone who wants to learn more about watersheds and how to take effective action to protect them. Bill takes us on a zany journey through a watershed, outlines the three-step process of becoming a Streamkeeper, and discusses concepts such as the hydrologic cycle, insects as indicators of stream health, and the politics of stream care. Real Streamkeepers show how to make a difference. People of all ages will be encouraged to investigate, monitor and take actions in their own watersheds. (25-minutes)

Adopting A Stream
A Northwest Handbook
Steve Yates
Illustrations by Sandra Noel

WRITTEN as a detailed, but easy-to-read overview of streamlife in the Pacific Northwest, *Adopting-A-Stream* applies to streams in all temperate climates. This comprehensive book tells how you or your school, community or club can protect and restore a nearby creek--and in the process learn about biology, ecology, economics, and the effects of watershed activities on our streams. For teachers, conservationists, fishermen, and lovers of the outdoors everywhere. (126 pages. Paperback)

Adopting A Wetland
A Northwest Guide
Steve Yates
Illustrations by Sandra Noel

A GREAT ADDITION to *Adopting A Stream* and an ideal resource for schools, community groups and individuals interested in restoring and/or protecting their neighborhood wetland areas. Information is in simple terms. You will learn about marsh life, wetland types and identification, their values and benefits, plants, mitigation and legislative issues. Appendices on wetland plants, wildlife, scientific classification and a basic observations checklist. (39 pages. Paperback)

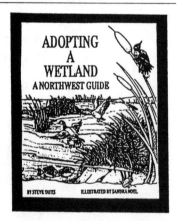

Salmon Come Back Poster

Designed in honor of the disappearing Snake River sockeye, this limited edition, strikingly beautiful full color poster will only be available while current supplies last. Created for Adopt-A-Stream by artist Sandra Noel. Suitable for framing. (18"x 24")

ORDER FORM FOR ADOPT-A-STREAM PUBLICATIONS

	Number ordered	Price Each	Total

Streamkeeper's Field Guide: Watershed Inventory and Stream Monitoring Methods.
ISBN 0-9652109-0-1

$29.95 +4.00 Shipping/handling.
WA residents add $2.37 sales tax.   _____ X _____ =_____

The Streamkeeper

$19.95 +3.00 Shipping/handling
WA residents add $1.58 sales tax   _____ X _____ =_____

Adopting A Stream: A Northwest Handbook
ISBN 0-295-96796-X

$14.95 +$2.00 Shipping/handling
WA residents add $1.18 sales tax   _____ X _____ =_____

Adopting A Wetland: A Northwest Guide
$5.00 + $2.00 Shipping/handling
WA residents add $.40 sales tax   _____ X _____ =_____

Salmon Come Back Poster
$9.95 + $2.00 Shipping/handling
WA residents add $.79 sales tax   _____ X _____ =_____

Total Order $_____

Total Shipping/handling $_____

Total WA sales tax $_____

GRAND TOTAL $_____
PAY THIS AMOUNT

Name_____

Address_____

City_____State_____Zip_____

The Adopt-A-Stream Foundation • 600-128th Street SE • Everett, WA 98208• phone (206) 316-8592